SMALL-SCALE TEXTILES

PLANT FIBRE PROCESSING

Other books in this series:

SMALL-SCALE TEXTILES

PLANT FIBRE PROCESSING

A handbook

Cyril Jarman

Intermediate Technology Publications 1998

Practical Action Publishing Ltd
27a Albert Street, Rugby, CV21 2SG, Warwickshire, UK
www.practicalactionpublishing.org

© Intermediate Technology Publications 1998

First published 1998\Digitised 2013

ISBN 10: 1 85339 385 1
ISBN 13: 9781853393853
ISBN Library Ebook: 9781780442990
Book DOI: http://dx.doi.org/10.3362/9781780442990

A catalogue record for this book is available from the British Library.

The authors, contributors and/or editors have asserted their rights under the
Copyright Designs and Patents Act 1988 to be identified as authors of their respective
contributions.

Since 1974, Practical Action Publishing has published and disseminated books and
information in support of international development work throughout the world.
Practical Action Publishing is a trading name of Practical Action Publishing Ltd
(Company Reg. No. 1159018), the wholly owned publishing company of Practical
Action. Practical Action Publishing trades only in support of its parent charity
objectives and any profits are covenanted back to Practical Action (Charity Reg. No.
247257, Group VAT Registration No. 880 9924 76).

CONTENTS

PREFACE

Almost all fabrics, whether woven or knitted, are produced from spun thread. However, a broad range of tasks must be undertaken before the spinning operation is reached.

The operations described in this handbook are those which will need to be considered when producing a fabric using plant fibres. The use of this information should improve the quality of the end product and thus make a positive contribution to the welfare of those involved in this sector of the textile industry.

This handbook is therefore concerned with the processes by which a wide variety of plant fibres are extracted, but it also provides background information on the growing areas, soil and climatic requirements and methods of harvesting the raw materials.

Cyril Jarman

FOREWORD

This handbook is one of a series dealing with small-scale textile production, from raw materials to finished products. Each handbook sets out to give some of the options available to existing or potential producers, where the aims could be to create employment or to sustain existing textile production. The ultimate goal is to assist in generating incomes for the rural poor in developing countries.

Needless to say this slim volume does not pretend to be comprehensive. The number of plants in the world which produce fibres and the number of technologies with which to process them, and which are already available, are numerous. This handbook is intended as an introduction which might stimulate further enquiry.

I am pleased to have had the opportunity of working with Cyril Jarman on this handbook. Cyril Jarman is the expert in the field of fibre extraction from plants, and together with my contribution of Chapter 5, Planning for production, we hope to identify some appropriate solutions to particular development problems. We are grateful for the assistance of Hew Prendergast and James Morley at the Centre of Economic Botany at Kew, who provided some of the information in Appendix 1.

The series of small-scale textile handbooks forms part of the process of identifying the need, recognizing the problems, and developing strategies to alleviate the crisis of unemployment and underemployment in the South.

Intermediate Technology offers consultancy expertise and technical enquiry services.

Martin Hardingham
Intermediate Technology, UK

1. INTRODUCTION TO PLANT FIBRES

This handbook contains information on fibres which are extracted from plants — often referred to as vegetable fibres. It encompasses not only those fibres such as cotton, sisal and jute which are traded on a massive scale, but also less well-known fibres which are nevertheless of great value to small communities where they are used for both practical and decorative purposes. With a few exceptions, the fibres have to be spun into a yarn before they can be made into useful articles such as twines, ropes and fabrics.

After harvesting, some fibres may require only to be cleaned, 'opened' and drawn out into a sliver, after which they are ready for spinning. This is the case with cotton, in which the fibres are stripped from the seeds to which they are adhering by a process known as 'ginning'. Other types of fibre are contained within the tissues of the stems of dicotyledonous plants and leaves of monocotyledonous plants or, in the case of coconut fibre, from the 'fruit' of the palm. The term 'fruit' is used here in its botanical rather than its everyday use, and refers to the structure which develops from the ovary and its contents after fertilization, also including structures formed from other parts of the flower or axis. Fibres first have to be extracted from these tissues.

This handbook contains descriptions of harvesting, extraction and processing, as well as background information on the growth, soil and climatic requirements for the plant fibres listed below.

TYPES OF PLANT FIBRE

Fruit fibres

Cotton	*Gossypium* spp.
Coir (coconut fibre)	*Cocos nucifera*

Stem fibres

Jute	*Corchorus capsularis* and *C. olitorius*
Kenaf	*Hibiscus cannabinus*
Flax	*Linum usitatissimum*
Ramie	*Boehmeria nivea*
Hemp	*Cannabis sativa*
Sunn fibre	*Crotalaria juncea*
Himalayan/Nilgiri nettle	*Girardinia diversifolia*

Leaf fibres

Sisal	*Agave sisalana*
Henequen	*A. fourcroydes*
Maguey	*A. cantala*, *A. atrovirens* and others
Pineapple	*Ananas comosus*
Abaca	*Musa textilis*
Fique	*Furcraea* spp.

GROWING AND HARVESTING

Fruit fibres

Cotton

Cotton has a world production of around 18 million tonnes and is one of the most important fibres used in large-scale textile manufacture. The main processors of cotton are the USA, the former USSR, China and India. Other important producers are East Africa, Egypt and the Sudan (see Appendix 1).

Two cultivated species of cotton, *Gossypium herbaceum* and *G. arboreum*, are grown in substantial quantities in India. *G. arboreum* is a perennial with short hairs, also grown by small-scale farmers. The 'upland' varieties of *G. hirsutum* provide the bulk of the cotton in the USA, the former USSR, China, South America and most of Africa. *G. barbadense* is less important and contributes only one-third of the world's cotton. However, both the 'Sea Island' and 'Egyptian' cultivars of this species yield fibres that are particularly long and lustrous. While the best-known by product of the cotton plant is the lint, the hairs which grow on the seed coat yield another very valuable product in the form of oil, which makes up 20 per cent of the seed (see *Illustration 1*). Cotton seed oil is used for cooking, margarine and soap. Cotton seed cake, the residue after milling, is a valuable animal feed. Cotton tolerates droughts better that most other annual crops, but the water requirements vary at different stages of growth, and very high rainfall is undesirable after the bolls (seed-heads) have opened. Cotton can be grown on a wide range of soils if there is adequate drainage.

Illustration 1 Open cotton boll (fruit fibre)

Cotton picking is highly labour intensive, and on a large scale is often carried out by machine. In many parts of the world, however, picking is carried out by hand. Since cotton must be picked at weekly intervals to prevent discoloration of the lint in the field, it is a very laborious task: smallholders average only 9 kg per day.

Coir (coconut fibre)

Coconut fibre is obtained from the husk of the fruit of the coconut palm, *Cocos nucifera*. The coconut palm is one of the most useful plants grown by man. Among its valuable products are copra (dried coconut flesh) and oil, as well as building materials and food products. *Illustration 2* shows a coconut in cross-section. Covering the nut is a tough, leathery layer called the exocarp. Beneath this is a mass of fibrous tissue. From this the fibre is extracted.

The origin of the coconut palm is unknown, but it is now widely distributed around the tropics and sub-tropics. It grows best in rich, loamy soil with good drainage and an underground flow of water, but will tolerate saline and sandy soils. For optimum production of nuts the plant needs a well-distributed rainfall of about 1250 mm per annum or more. Palms suffering from lack of water yield fewer nuts. There are three main types of palm: 'tall', 'dwarf' and 'hybrid'. A reasonably good annual yield of the tall variety is around 60 nuts per tree; this is equivalent to about 10 000 nuts per hectare. Coconuts vary greatly in size. Those from Sri Lanka, for example, are very large, yielding the long fibres essential for making brushes.

If the nuts are going to be used to produce 'white' fibre (for the production of yarn), they are harvested when they are still green, after about nine months' growth. Nuts which are intended for brown fibre (to be used in brushes and mattresses) are harvested when they are brown, after twelve months or more of growth. Brown nuts give the maximum yield of white flesh which is either made into desiccated coconut or dried to form copra. Green nuts yield only a small amount of white meat which is used locally in cooking.

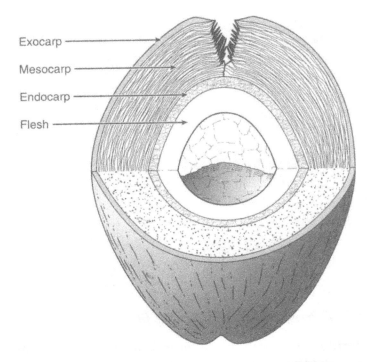

Illustration 2 Coconut fruit — longitudinal view

The 'tall' variety of coconut palm can grow to a height of 50–70 ft (15–20 m), making harvesting difficult. In some parts of the world the nuts are collected after 'free fall'. This can be dangerous, as the heavy nuts have potential to do great damage, and they can easily be overlooked lying in the undergrowth. Therefore, in many countries, people climb the trees to harvest the nuts. A group of three people with two climbers can pick the nuts of about 80 trees in an eight-hour day (five nuts per tree, so about 400 nuts per 0.5 hectare). Using a pole with a knife, three people can harvest and gather nuts from two hectares (1560 nuts) in a ten-hour day. A trained monkey is used in some countries; assuming each (tall variety) tree yields five nuts, a monkey will harvest 350 nuts per day.

Stem fibres

Jute

Jute, with an annual world production of around 3.5 million tonnes, is the second most important vegetable fibre after cotton. Together with kenaf and roselle (see below), it has long been of major importance to agriculture in many tropical and sub-tropical countries, particularly Bangladesh and India.

Jute is obtained from the stem of the jute plant, *Corchorus olitorius* ('Brown', 'Tossa' or 'Daisy' jute), and *Corchorus capsularis* ('White jute'). The genus *Corchorus* belongs to the lime family or Tilleaceae.

White jute grows to a height of 11.9 ft (3.7 m) while Brown jute reaches 14.7 ft (4.5 m). The cylindrical stem has a diameter of from 0.5 to 1.5 in (1.25 to 4.0 cm) at the butt end, tapering towards the tip. Brown jute has branches, not always found with White jute.

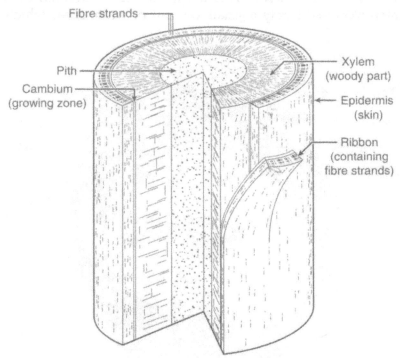

Illustration 3 Three-dimensional view of jute stem (stem fibre)

Jute thrives only in fertile soil where temperatures are high and rainfall is 20 cm or more each month of the growing season. White jute endures flooding, as experienced in the delta lands of the Ganges and Bramaputra rivers. Brown jute is grown only in higher areas not subject to flooding. The soil must be well cultivated before seed sowing.

When harvested, the plants are cut near the ground, using a sickle-type knife, tied in bundles and left in the field for two or three days. It is recommended that the jute crop is harvested at the 'small pod' stage to obtain the best combination of yield and quality. After harvesting, the plants are laid in rows on the ground. This facilitates heating, causing the leaves to drop from the stem, reducing the weight by as much as a third, making transportation to the retting place easier. Defoliated stems ret more evenly and produce cleaner fibre.

Kenaf, roselle and urena

Kenaf from *Hibiscus cannabinus* and roselle from *H. sabdariffa* var. *altissima* yield fibre similar in appearance to jute and are extensively used as jute substitutes for sacks and hessian. They are both less demanding than jute in their soil and climatic requirements.

Kenaf will grow better in drier conditions and poorer soils than jute and most other bast fibre plants, the principal requirement being that the soil possesses good drainage (although the plant will tolerate flooding in the late stages of growth). The main kenaf growing areas are within a range of latitude from 16°S to 41°N with a mean relative humidity of 68–82 per cent. The mean temperature during the growing season ranges from 22 to 30°C, and the mean monthly rainfall during the growing season ranges between 10 and 329 mm. All the important jute substitutes, namely kenaf, roselle and urena, belong to the mallow family (Malvaceae).

Roselle is even more adaptable than kenaf in that it will thrive in poorer areas. However, since it is a highly depleting crop (it draws the goodness out of the soil), adequate crop rotation and fertilization must be carried out to maintain high soil fertility.

The plant tolerates both highly acid and moderately alkaline soils. The major roselle growing areas cover a range of latitude from 7° 0"S in Java to 23° 46"N in Bangladesh. Roselle prefers a high humidity, 70–86 per cent during the growing season, with a mean temperature range of 25–30°C and a mean monthly rainfall of between 135 and 274 mm.

Urena, from *Urena lobata*, appears to thrive on light, deep, friable soil, rich in potassium. Almost all commercial plantings of urena are grown at latitudes ranging from 0 to 20°N or S. The plant prefers a high relative humidity ranging from 73 to 85 per cent with an average temperature of 21–27°C and rainfall ranging from 160 to 205 mm per month during the growing season.

Kenaf can be harvested by hand in a similar manner to jute, although when grown in large plantations it is harvested mechanically. In most areas where Roselle is grown, especially in Thailand, the crop is pulled if the soil is sandy. Urena plants are cut about 20 cm above ground, since the stem base contains much lignin (see Glossary) and does not ret easily.

Flax

Flax, from *Linum usitatissimum*, is an annual herbaceous plant belonging to the Linaceae family. It is a plant of temperate areas grown mainly in the former Soviet Union, Poland, France, Belgium and the Netherlands. It requires an abundance of readily available plant nutrients and a neutral or slightly acid soil. It is cultivated for fibre and for oil-bearing seed. The varieties grown for fibre have slender, erect stems ranging from 27.5 to 47.2 in (70 to 120 cm) in height and 0.06 to 0.08 in (1.5 to 2.0 mm) in diameter.

The crop is harvested when the stems have turned from green to yellow: usually about one month from the appearance of the first flowers. Flax, unlike other fibre crops, is never harvested by cutting, but the stalks are pulled from the ground. This avoids the loss in fibre length from stubble left in the field. Pulling flax by hand is very laborious work, requiring as much as twenty person-days to harvest one hectare. Mechanized harvesters can pull one hectare per day.

Ramie

Ramie is cultivated mainly in China, the Philippines, and Brazil. The fibre is obtained from the stems of plants belonging to the genus *Boehmeria* of the Urticaceae or nettle family. Ramie is a hardy perennial shrub in which stems (or canes) with little or no branching are sent up from an underground root stock (rhizome). The stems grow to between 3.9 and 7.7 ft (1.2–2.4 m).

The first crop after planting the rhizomes is very uneven in height and usually branched, making it unsuitable for fibre extraction. The second crop is usually very robust but with little fibre. In temperate areas three crops are harvested annually. The optimum harvest period is only ten days, after which the plant becomes woody and therefore difficult to process.

Ramie is usually harvested by hand. Although it can be cut with a mower, unless previously removed, the leaves tend to tangle around the cut plant and make manual separation and uniform butting (see Glossary) difficult.

Hemp

Hemp is an annual plant known botanically as *Cannabis sativa*, belonging to the family Cannabidaceae. When mature, hemp develops a rigid woody stem, ranging in height from 3.3 to 16.5 ft (1 to 5 m). Hemp grows best where the temperature range between 13 and 22°C. It requires a plentiful supply of moisture during its growing season, ideally 125 mm per month. It grows best in a deep, well-drained clay-loam soil containing considerable organic matter. Hemp is dioecious — that is, it carries male and female flowers on different plants. The male plants are taller than the female plants. Monoecious (male and female flowers on the same plant) varieties have been developed by breeding.

In many countries hemp is harvested by hand. A skilled harvester is able to cut and lay hemp for 0.3 hectare per day. There are, however, tractor-drawn harvesters, used for a range of crop harvesting, which can be modified for use with hemp.

Sunn fibre

Sunn fibre from *Crotalaria juncea*, a leguminous crop, is usually considered to be a native of India. It is however grown in both temperate and tropical areas of the world for several purposes: as a bast fibre, green manure and sometimes forage crop for livestock. It is also called Indian hemp, Bombay hemp, Madras hemp and Jubblepore hemp. The sunn fibre plant usually has many branches, but in dense stands it grows as a single stem that can reach a height of 5 to 12 ft (1.5–3.6 m) in height or more. Sunn fibre grows on almost any type of soil that is free from waterlogging, but it prefers a well-drained light loam.

Practically all sunn fibre is harvested by hand. As with jute, plants are cut or pulled and left in the field for two days to cause the leaves to drop. Harvesting is later if the fibre is intended for paper making.

Himalayan or Nilgiri nettle (allo)

The Himalayan or Nilgiri nettle, *Girardinia diversifolia*, known locally as allo, grows under forest cover at high altitudes in the middle hills across Nepal. The plant grows to a height of over 9 ft (3 m), with perennial roots, a fibrous stem which grows up to 1.5 in (4 cm) in diameter at the base, with large-lobed serrated leaves. Both stems and leaves are covered with long, thorn-like stinging hairs.

The plant has a variety of traditional uses. Bark fibre from the stem is extracted, processed and spun to produce a yarn. The young, tender shoots are used as a vegetable and medicinal herb, and also for pig feed. Allo yarn is used to make fishing nets, or woven into the narrow head bands used for portering. It is also extensively used for weaving sacks, small bags and mats, and attractive patterned cloth.

The mature thick stems are cut with a kukri or sickle. The outer bark which contains the fibre is then stripped off by hand. One person can harvest about 37.5 kg in a day. The green bark is taken home and part may be processed immediately, while the remainder is dried. When dried, it can be stored indefinitely and then soaked in water before processing.

Fibres from leaf sheaths

Sisal

This the most important of the leaf fibres, extracted from a type of agave, *Agave sisalana* Perrine. There are many different species of agave, all originating in South America where the fibre has many traditional uses. *A. sisalana* was introduced into East Africa, where commercial plantations were established in the early 1900s. East Africa is still a main producer of sisal, but has for many years been rivalled by Brazil. In both countries a substantial quantity of fibre has been produced from high-yielding hybrids.

Sisal is tolerant of a wide variety of soils, providing they are friable, freely draining and not too acid or low in nutrients. Optimum rainfall for sisal is between 1200 and 1800 mm, and should be well distributed throughout the season. Sisal is often considered to be an arid zone plant, but where rainfall is less than 760 mm there will be erratic growth, or even no growth for long periods. A hot and equable climate suits sisal best, with maximum temperatures between 27 and 32°C. Temperatures should not fall below 16°C.

Sisal produces rhizomes (underground stems) from buds at the base of the plant. They grow upwards at some distance from the plant to produce a new plant known as a 'sucker'. A plant may produce about twenty such suckers in its lifetime. Sisal may be

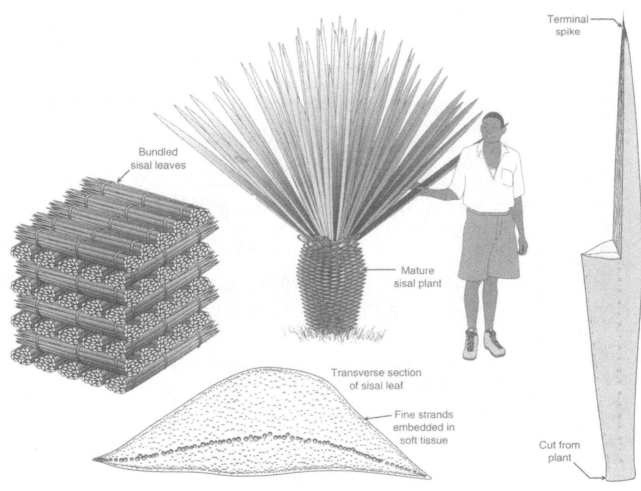

Illustration 4 Cutting sisal leaves

7

propagated by means of these suckers, and they are frequently used as planting material.

Sisal flowers only once during a lifetime of seven to twelve years, after which it dies. It is rare for the seeds to set, but miniature sisal plants or 'bulbils' are carried on the flowering head (inflorescence), and when mature these fall to the ground. They are sometimes used instead of suckers as planting material.

The sisal leaf is attached directly to the stem or bole. It may attain a length of two metres, but is commonly about 120 cm long. At the tip of the leaf is a large spine. Unlike henequen (see below) there are no spines along the edges of the leaf.

Harvesting consists of first cutting off the leaf from the plant, removing the tip and any unhealthy portions of the plant, and placing the cut leaves lengthwise between the rows. The second part of harvesting consists of bundling, binding and transporting the leaves to the ends of the rows. They are then collected and taken to the decorticator.

When plants are grown on fertile soil without severe or prolonged drought, the first harvest of the basal leaves may be made two or three years after planting. Cutting (*Illustration 4* on page 7) may be repeated at six-, nine- or twelve-month intervals, in each case leaving about 20–25 leaves uncut for further growth. Over-cutting seriously reduces yields.

Henequen

Henequen, or Yucatan sisal, is obtained from *Agave fourcroydes* Lem., and is grown on a large scale in Mexico, mainly on limestone plateau in the Yucatan peninsula. The fibre yielded is similar to sisal, used mainly for agricultural twine but also for cordage, upholstery padding and floor coverings. The annual rainfall of the Yucatan growing region varies from 500–600 mm to about 1100 mm. The mean temperature is 26°C. The life cycle of the plant is 25 years, twice as long as that of *A. sisalana*. It takes eight years before it is mature enough for the first harvest. About 12–14 leaves are left on the plant after cutting.

Maguey

The species usually referred to as maguey or 'pulque' (*Agave cantala* Roxb. ex Salm-Dyck, *A. salmiana* Otto ex Salm, *A. mapisaga* Trel.; *A. atrovirens* Karw. ex Salm; *A. ferox* Koch; *A. hookeri Jacobi* and *A. americana* L.) grow in the highlands of central and north-central Mexico. After fermentation, pulque is used as a beverage.

Once the plant has been prepared for sap ('aguamiel') production, it will produce sap continuously for three to six months. If, however, it produces for over four months, the plant becomes dry and leathery and fit only for fuel. If sap production is carried out for a shorter period, the maguey leaves remain green and pliable and the fibre can still be extracted by scraping away the flesh, so it is possible to use a mature plant for production of both sap and fibre.

Generally, larger maguey leaves contain relatively coarse fibres. The finest fibres come from the smaller leaves, usually whitish in colour, which cluster at the centre of the plant. These are particularly suitable for the production of fine cloth.

Pineapple fibre

Pineapple fibre is obtained from the leaves of the pineapple plant, *Ananas comosus* L. Smooth Cayenne is a widely grown variety, but there are many others grown for their fruit. Pineapple is grown in sub-tropical countries, including the Philippines, Taiwan, Brazil, Hawaii, India, Indonesia, and the West Indies.

The sword-shaped leaves arise from a stem about 2 ft (60 mm) high. The inflorescence appears at the tip about 12–15 months after planting, depending on the type of planting material used. Smooth Cayenne grows leaves on average 23 in (58 cm) long, 1.8 in (4.7 cm) wide and about 0.1 in (2 mm) thick.

In the Philippines a non-fruiting variety, Spanish Red, is grown solely for fibre. This is hand-woven into the sheer fabric used in 'Barong Tagalog', the traditional formal shirt for men. This variety has a higher yield of fibre and the leaves are longer than the varieties grown for fruit. Longer leaves are produced if the plants are more closely spaced and also grown in the shade.

A ratoon crop, developed from suckers, is harvested 14 months after the main crop.

Abaca

This is also known as 'Manila hemp'. It is obtained from *Musa textilis* Née. Abaca is a perennial growing from short rootstocks. A mature abaca plant or 'mat' consists of a group of stalks ranging from 6 to 15 ft (1.8 to 4.6 m) high. These stalks (pseudostems) are made up of a central core which is encircled by overlapping leaf sheaths, each bearing a frond 3-6 ft (0.9–1.8 m) long and about 12 in (30 cm) wide. Abaca requires a warm, moist tropical climate and a deep, well-drained, fertile soil. There are some 200 varieties of abaca in the Philippines, where it is grown mainly on small, multi-crop farms. The remaining production is almost entirely from Ecuador, where it is grown on large estates.

Fique

Fique, also known as cabuya, is obtained from species of *Furcraea*, particularly *F. macrophylla* Baker and *F. cabuya* Trel. It is produced on a much smaller scale than sisal, and the fibre is extracted from the leaves using a mobile decorticator. The optimum temperature for growing fique is between 18 and 24°C. In Colombia it is used for making coffee sacks.

Fique leaves may be cut annually, six-monthly or four-monthly, depending on the rate of growth of the plant. Only mature leaves are cut, namely those which make an angle of not less than 40 degrees to the main plant.

2. FIBRE EXTRACTION AND PROCESSING

Some fibres are embedded within the tissues of the plant and this tissue has to be removed to obtain the fibre. Other fibres, such as cotton and kapok, need only be removed from the seed case. Fibres can be extracted from tissues either by beating and/or scraping the fresh leaves or by first softening the tissues by retting. For some fibres retting alone is not sufficient and a further chemical treatment has to be carried out. An example of this is the degumming of ramie using alkaline solutions.

FRUIT FIBRES

Cotton

Cotton fresh from the bale has the appearance of 'cotton wool' mixed with pieces of dead leaf and other debris. Hand-picked cotton is, however, cleaner than machine-picked. Samples from different sources vary greatly in cleanliness, staple length and colour.

The first stage in processing the cotton bolls into fibre suitable for spinning is the removal of the cotton seeds by ginning. In West Africa the following method is used. A few bolls at a time are placed on a block of wood, or in some areas a flat stone, and their seeds are squeezed out by rolling an iron, or sometimes a wooden rod, over them. The mechanical ginning process was invented by Eli Whitney in 1793 and the same principle is applied today (*Illustration 5*).

The surface fibres are caught on a battery of toothed discs or combs and pulled through slots too narrow for passage of the seed. After the fibres have been torn from the seed, they are brushed loose or blown free from the disk or comb. The GKGM (Gujarat Khadi Gramodyog Mandal) has developed small-scale gins for use by craft co-operatives.

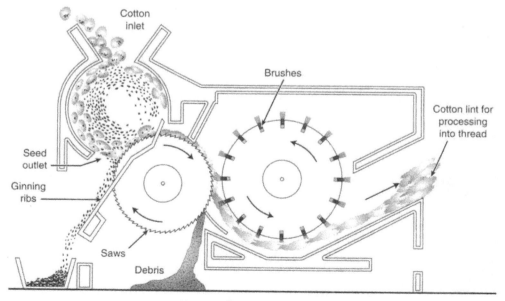

Illustration 5 Cotton gin

Table 1. Preparatory processes for spinning

Cotton	Flax and hemp	Jute	Sisal and abaca
Bale breaker Breaks apart large compressed pieces	**Breaker** 1st stage in separating fibre from woody core. Breaks up woody core	**Bale opener** Softens fibre from compressed bale by passing through crushing rollers	**Hackler** Removes tow and extraneous materials, combs ends
Opener Further breaking apart and separation of heavy debris		**Softener** Softens and batches fibre by passing through large number of fluted rollers	**Breaker (Good's machine)** Combs, drafts and converts fibre bundles into a continuous sliver
Picker Further cleaning and conversion to a sheet of fibre tufts (lap)	**Hackler** 2nd stage in separating fibre. Combs fibres into finer form, removes tow and extraneous materials		
	Spreader Converts fibre bundles into a continuous sliver	**Spreader** Combing action converts fibre bundles into continuous sliver	**Spreader (Good's machine)** Combing, doubling and drafting
Carder Cleaning and parallelizing of fibres	**Carder** Cleans fibres and converts to a mat of parallel fibres when spreader is not used	**Carders (breaker and finisher)** Cleans fibres and converts to a mat of parallel fibres	
Drawing frame Doubling and drafting	**Drawing frame** Doubling and drafting	**Drawing frames** (3 stages) Doubling and drafting	**Drawing frame** Doubling and drafting
Roving frame Drafting continued, slightly spun, wound on bobbin	**Roving frame** Drafting continued, slightly spun, wound on bobbin	**Roving frame** Drafting continued, slightly spun, wound on bobbin	**Finisher** Drafting to size for spinning

Cotton is usually transported from the gin compressed into bales weighing 400 lb (about 180 kg). On arrival at the mill the bales are opened using a bale-breaker. The subsequent pre-spinning processes are shown in *Table 1*.

Coir (coconut fibre)

In practically all coconut-producing areas, the husk is removed by impaling the coconut on a sharp iron spike. The coconut is taken in both hands, the operator brings it sharply down on the spike and, with a combination of cutting and twisting actions, the husk is removed. The method requires a high degree of skill and strength by the worker (dehusker) if a reasonable level of output is to be achieved.

In India, using the simple device shown in *Illustration 6*, a skilled worker can dehusk 1500 to 2000 nuts in a working day. In Sri Lanka young men dehusk on average about 1500 nuts a day. In the Philippines the nuts are dehusked using a transportable wooden horse fitted with a vertical spike. The husks are removed from the nuts in three or four segments, depending on the size of the nut. One man using this device can dehusk about 1000 husks in a ten-hour day.

A number of devices have been developed to assist in the manual dehusking operation, attempting to improve upon the traditional spike method. None of these has proved satisfactory, and the spike remains the main method of dehusking.

Fibre from immature green husks is referred to in the trade as 'white' fibre to distinguish it from the 'brown' fibre from brown mature husks. However 'white' fibre varies in colour from cream to grey depending on its origin. Brown fibre is generally golden brown in colour but is lighter after it has been bleached with sulphur dioxide.

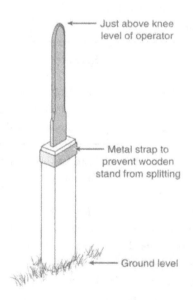

Illustration 6 Coconut dehusking spike

The fibres contained in the husk are obtained by beating and washing, but first the husk segments have to be softened. This is achieved by a process known as retting. This is a process similar to the rotting which takes place naturally, in which plant material is broken down into compost-like matter. The difference between retting and rotting is that in retting, the process is carefully controlled and ended as soon as the non-fibrous matter has been softened, but before the fibres themselves have been attacked.

The retting process can be divided into two stages. In the first or 'physical' stage, the water-soluble constituents dissolve. Some of the compounds such as the polyphenolics are lethal to bacteria and these must be removed before the next 'biological phase' can commence. A variety of micro-organism develops which metabolizes the extracted substances and creates a suitable environment for a different group of micro-organisms, which decomposes the binding material of the tissues, collectively referred to as pectic substances. This micro-biological activity leads to the production of organic acids and gas, accompanied by a rise in temperature of the husks. During the biological stage the anaerobic organisms decompose the pectins in the middle lamella of the parenchymatous tissues, and as they dissolve the fibre bundles separate. Most of the fibres become loosened after about six months, but about ten months are required for the full loosening.

Methods of retting
The method of retting varies from place to place. In 'pit retting' which is commonly used, the husks are buried in basin-shaped pits dug on the banks of backwaters; the necessary water movement is provided by the ebb and flow of water at the top and the percolation of water from the subsoil below. Alternatively the pits are provided with channels to allow for the flow of water. The bottoms of pits are covered with sand and the sides are lined with coconut leaves. After filling, the pits are covered with coconut leaves and weighted down with mud to prevent the husks from floating. Each pit can hold as many as 20 000 to 30 000 husks.

The best quality of coir is produced by placing husks (up to 10 000) in coir nets called 'vallies' and transferring them to brackish waters subject to tidal action. The husks are immersed by weighting with mud and stones. The retting period ranges from six to nine months, after which the mass is removed for further processing. After retting, the softened husks are removed from the retting water, washed, and finally squeezed in water to get rid of the mud and smell.

The washed husk segments are then taken to beating places in the shade, on the banks of the lagoons. The fibre next to the hard inner shell is short and is removed first, together with the tough exocarp. The fibrous material within is kept for beating out into the longer yarn or mat fibre. These pieces of fibrous tissue are placed one by one on a piece of wood and beaten by means of a strong, round, wooden rod until all the pith is removed. After it has been well beaten, softened and shaken, the fibre is washed and spread out by hand to remove any clogged, pithy matter remaining.

The retted husk has a foul smell and beating it by hand is unpleasant. Moreover it is a low-paid job and the output per worker is low. It is possible for a woman worker to be able, on average, to beat 100 husks in a ten-hour day. To overcome these difficulties, husk beating machines were introduced in various parts of Kerala from the mid 1960s onward. These motor-powered machines can beat 10 000 husks per eight-hour day and can also, it is claimed, beat husks that have been retted for shorter periods. These machines have not however gained acceptance by coir workers.

Teasing, winnowing and baling
Fibre to be used for door mats is teased by passing it through a picking machine, whilst that to be used for spinning (especially for the highest quality yarns) is cleaned more thoroughly by passing it through a winnowing machine.

Hanks of bleached and unbleached yarn are bound into bundles for transport to the shipper's stores. They are, after inspection, baled in a high-density press.

STEM FIBRES

Jute

Retting is almost always carried out by immersing the stems in water, when the bacteria on the stem will attack the plant tissues surrounding the fibres, softening them sufficiently so that they can be washed away leaving the fibres unaffected. Retting can take between 5 and 22 days depending on temperature and pH. At the optimum temperature of 35°C, retting takes about 7 days.

The stems are formed into bundles before being immersed in pools, canals, slow-moving streams or ponds, at a depth of 3.2 to 6.4 ft (1–2 m). Retting is complete when the bast separates easily from the inner woody core of the stem. The bundles must be removed and the fibre stripped from the stem before it over-rets, since over-retted fibre has poor strength and hence little value.

The retted stems are first gently beaten at the base with a mallet to loosen the fibre. The woody core is then broken near the base and the broken pieces discarded. The fibre is stripped from the core (*Illustration 7*). Afterwards, adhering pieces of bark and broken stick are removed by lashing the stripped fibre on the surface of the water. Finally the water is wrung out of the fibre. The washed fibre is usually dried on horizontal wires or poles. When dry, the fibres are made up into hanks or bundles. Retting is the most important step in the production of good quality jute. It is a process which requires considerable experience in deciding the right time to remove the stems from the retting water. More fibre is downgraded through under- or over-retting than for any other reason. The presence of hard root in the fibre is a main cause for downgrading; however the amount can be reduced by softening the root end using urea solution, or malleting.

Immature jute plants

Illustration 7 Stripping fibre from woody core of jute

Ribboning

An alternative method to the retting of whole stems is to ret only the outer fibre containing part of the stem. The operation, known as 'ribboning', produces strips of fibre containing bast. It has been used in some countries for many years but is now receiving wider interest. Ribboning can be carried out using either simple hand-operated devices or by motor-powered ribboning machines. Either fresh or dried ribbons can be retted.

The simplest device is a stout vertical pole driven into the ground, protruding about one metre. The operator loosens about 6 in (15 cm) of raw bast from the butt-end of the stem, and pulls it around the pole towards himself. The clean woody core is propelled beyond the pole and the operator is left with a bast fibre ribbon.

The bicycle hub ribboner (*Illustration 8*) is not quite so basic. It is used in South-east Asia and consists of a horizontal board fitted with the hub from a bicycle wheel. Its use reduces the physical effort needed to ribbon kenaf stems.

Both these simple ribboners require a high labour input, and in countries where labour is not available, mechanical ribboners have been used. Some of these are small while others designed for use on plantations are large and very expensive with a high output.

The Alvan Blanch ribboner (*Illustration 9*) is based on an existing small-scale rice thresher. It consists basically of a drum 30 cm in diameter to which are attached three beater bars. Stems are pushed, butt-end first, along a feed platform and through a slot in the cowl. Above the slot is an adjustable anvil bar. The bars on the revolving drum strike the stem upwards and cause the inner woody core to be broken and then beaten out. The separated fibre ribbons are carried over the drum to trail on the discharge side where they can be lifted off with a stick.

The **advantages** of ribboning are:

- ❏ Less water is required for retting
- ❏ Valuable plant nutrients are returned to the soil
- ❏ There is less weight of material to transport
- ❏ Retting may be carried out at any time
- ❏ Fibre washing is easier.

The **disadvantages** of ribboning are:

- ❏ Additional labour is required for green stem stripping by hand
- ❏ Mechanical ribboning requires a high capital outlay on ribboners
- ❏ Retted fibre from ribbons tends to be inferior to that produced from retted whole stems.

The retted fibre is taken by the farmer to collecting centres where it is sorted and graded and then made into low-density or 'kutcha' bales, weighing from 60 to 150 kg, for transport to the local mills or to the high-density balers where they are made into 'pucca' bales weighing 180 kg and measuring 124 by 47 by 46 cm. On arrival at the mill the bales are opened. The subsequent pre-spinning operations are shown in *Table 1*.

Raw jute consists of stricks or bundles of fibres from one stem which are usually from 6 to 10 ft (1.83 to 3.05 m) in length. The fibre usually has pieces of bark adhering to it. 'White' jute varies in colour from creamy white to light reddish to straw colour and light grey. 'Tossa' jute varies from light golden to reddish but can also be light to medium grey.

Kenaf, roselle and urena

Kenaf, roselle and urena are retted in a similar manner to jute when grown by small-holders. When they are produced on a plantation scale, however, mechanized forms of retting are employed, using ribbons containing the fibre present in the outer part of the stem (see above).

Kenaf, roselle and urena are also processed in a similar way to jute. All three fibres are very similar in handle to jute, but urena is usually lighter coloured.

Kenaf stalk is pulled through the bench and the fibres are stripped against the bicycle hub

Illustration 8 Hand-ribboning kenaf stems using bicycle wheel hub

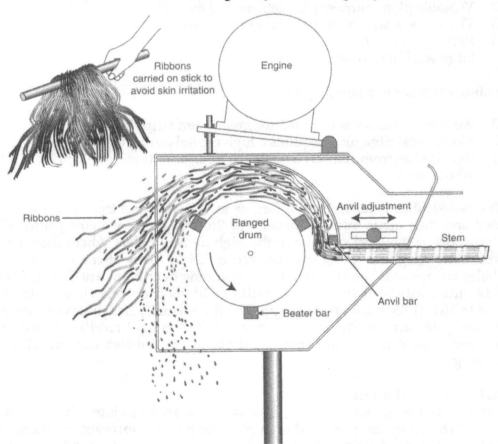

Illustration 9 Alvan Blanch ribboner

Flax

The retting of flax is carried out by micro-organisms in the same way as for jute and coconut fibre. However there are two basic methods: dew retting and water retting.

Dew retting is the easiest and cheapest of the two methods. Most of the fibre in Russia and Eastern Europe is dew retted, whereas in Western Europe water retting is preferred. Since dew retting is not an operation that can be controlled precisely, the quality of fibre produced is inferior to that produced by water retting. In dew retting freshly pulled straw is spread in the field in relatively thin layers. Rain, snow or dew causes fungi and bacteria to enter the stomata of the stems and attack the pectin binding the fibres together. Since there are so many variables in dew retting, there is no set length of time by which the retting will be complete.

Water retting in cold water is the most basic form of retting, and is carried out in rivers, pools and ditches. It is used by farmers who also process their flax by hand. The small, loosely bound bundles are submerged in water and weighted down with earth or stones to keep them submerged. At temperatures below 15°C there is little bacterial activity, so cold water retting can only be carried out during the summer months. The retter tests the straw frequently to determine if the fibre is separating from the woody core, a sign that the process has been completed. However, the retter has little control over the speed of the process, and nowadays the best flax fibre is produced by *warm water retting*, in well-equipped retteries that operate under controlled conditions to produce fine, uniform fibre. Typically a large Belgian retting establishment would have sixteen concrete retting tanks measuring 13 by 16 ft (about 4 by 5 m) in area and 8 ft (about 2.5 m) in depth, each capable of retting 100 tonnes of straw per ret or 6.25 tonnes per tank. The bundles of straw are placed in the tanks in two upright tiers, one upon the other. Cold water is added, and after eight hours the tank is drained. This serves the purpose of washing the stems, driving out the air in the straw and extracting the water-soluble constituents. This leaching process leads to a loss of 10 per cent of the original dry straw weight.

To re-initiate retting, fresh water at 25°C is added to the tank and warm water added from the bottom of the tank until a temperature of 28°C is reached. Water is added at intervals of 8–10 hours, leading to a complete change of water in the tank. This addition of water is necessary not only to maintain the optimum temperature but also to dilute the retting water. This becomes increasingly acid during the ret and inhibits the activity of the bacteria. The duration of the retting period is approximately four days. When flax straw is retted with warm water the principal retting organisms are anaerobic bacteria, as opposed to fungi in the dew retting process.

Once the retting is completed the straw bundles are rinsed, removed from the tanks, and taken back to the field where they are piled in open 'wigwam' fashion and allowed to dry completely before fibre extraction.

In Belgium it is customary to double-ret the better grades of straw. This procedure consists of bringing the first ret almost to completion in about two and a half days, draining the tank and rinsing the fibre with fresh water. The straw is then air-dried in the field and returned to the retting tank. A second short ret takes place at 34°C for one day only.

When the straw is completely dry it can be stored before the fibre is extracted — a process known as flax dressing. The early method of extracting the fibre was to beat the dry straw with a mallet to break up the shives, followed by beating and shaking the broken straw to remove most of the wood, and finally 'hackling' to align the fibres and remove the last shive portions and broken fibres. The process can be speeded up by using a flax breaker: a chopping device with two or three blades pivoted at one end.

Later, machines with fluted rollers (five pairs) were used, each progressively finer in depth and width, placed side by side horizontally. The stricks of flax straw were fed in to the coarser rollers and collected from the other end.

A further improvement in productivity was obtained by using the 'Flemish wheel', a powered unit consisting of a number of flexible wooded blades attached to a wheel. As the wheel revolved, the blades beat the broken straw held between them and a stationary board.

With the development of turbine scutchers about 50 years ago, a considerable increase in output was achieved. One person could clean 30–35 kg a day on a scutching wheel. This increased production by 200–350 per cent and produced fibre of much better quality with very little waste. The turbine scutcher is a mechanized combination of the multi-paired fluted roll breaker and automatically fed scutching wheels.

When dry retted straw is fed into the turbine scutcher, it is first broken repeatedly by the numerous pairs of partially intermeshing corrugated rolls. The broken partially shive-free fibre then passes to the scutcher section. The fibre is firmly gripped about two-thirds of the way from the root end, which is cleaned first. Then, by means of a transfer device, the cleaned portion is gripped while the upper third is cleaned by a second, offset, drum.

Most linen flax fibre is baled without pressure in hessian-covered bales that weigh 100 kg. On arrival at the mill the fibre is cut into a suitable length in a breaker. The subsequent pre-spinning operations are shown in *Table 1*.

The colour of water-retted flax varies from off-white to steel blue. Dew-retted flax may vary from light brown to grey-black, but the uniformity of colour is as important as the colour itself in judging the fibre.

An ideal length of fibre is 83 cm, not shorter than 70 cm and not longer than 90 cm. The presence of shives denotes either insufficient retting or faulty scutching.

Ramie

Ramie fibre is embedded in the cells of the bast which lies between the outer bark and the woody core of the stem. The fine, spinnable fibre is composed of long single cells fixed together in bundles by gums and pectins. Since these must be removed by degumming before the fibres can be spun, fibre extraction is a more complex process than with other bast fibres such as flax and jute.

Extraction of the fibre bundles is carried out in two stages. Firstly, the cortex, containing the bast and outer bark, is removed from the stem; this is sometimes called decortication and can be carried out by hand or machine. Secondly, the cortex is scraped to remove most of the outer bark and the parenchyma in the bast and some of the gums and pectins.

Extraction of the fibre bundles is followed by washing, drying and degumming before the ultimate fibres are obtained.

The degumming process is normally carried out using chemicals, although promising results have been obtained using microbial degumming.

In chemical degumming, hot alkali is used to dissolve the pectic substances which bind the ultimate fibres into bundles. Ramie mills tend to keep their processes secret, however, the main factors which affect the efficiency of chemical degumming are: concentration, pH, volume and temperature of the alkaline liquor, the duration of degumming and the freedom with which the liquor circulates through the fibre. Degumming can be carried out either below ('low temperature') or above atmospheric pressure ('high temperature'). High temperature methods produce stronger and cleaner fibre but yields are lower. Modern low-temperature or open vessel methods are simpler, cheaper and quicker than high pressure methods which they have replaced.

Many commercial degumming methods have been described, but the following method is claimed to be simple and cheap.

Fibre, in the form of ribbons which should be freshly decorticated, is given two successive boils for two hours each in lime water. It is then washed and rinsed in cold water. It is boiled for half an hour in a 1 per cent solution of carbonate of soda, and washed and rinsed in cold water. It is then treated for five hours in a cold solution of chlorinated lime, containing 1 per cent of chlorine (equivalent to about 10 g/l available chlorine), and washed and rinsed in cold water. The fibre is then dried, combed and carded.

Ramie comes on to the market in a variety of forms, namely:

China grass This consists of strips containing fibre cemented together by gums and outer bark. It contains 33–35 per cent gums. It is sometimes bleached.

Ribbons These consist of fibre, gum and bark. They are green when fresh but turn brown on storage.

Degummed fibre After removing the gums the fibres become separated and can be further processed to obtain maximum yield. Degummed fibre is white, silky and lustrous.

Tops These are obtained by combing the degummed fibres into a mat formed of parallel fibres.

Rovings These are produced by drawing the tops out to form a long thin strand or sliver.

Yarn The sliver or roving is twisted, so locking the fibres into a strand.

Hemp

As in the case of flax there are two methods of extracting the fibre.

Dew retting, although it produces the coarsest fibre, is more widely used since it can be mechanized and is cheaper than water retting. As in the case of flax the stalks are spread in the field in thin layers and turned over once or twice to give even retting. Retting usually takes four to five weeks, but if the weather is warm and humid, as little as one to two weeks. If it is not stopped at the right time retting will become rotting. The retter breaks several stems at once; if the wood separates easily from the fibre, the retting is complete.

Water retting is carried out in ponds and ditches. The stalks, sixty to ninety, are tied in bundles and immersed in the water, which must be at least 15°C. The rafts are weighted down with large stones. From the fifth day stem samples are taken to assess the degree of retting; the average retting time is 7–9 days. The bundles of retted stems are opened and a worker beats them on the surface of the water to wash them. The bundles are then retied, lifted out of the water and sun-dried in the field. When the stems are thoroughly dried they are stored.

After the retted stems have been taken out of storage the fibre is separated from the woody stem. There are two main methods of doing this.

The first method consists of hand-breaking, followed by hand hackling to remove the woody stem portions called 'hurds'. Hand-breaking produces the finest quality hemp with by far the lowest losses.

A simple wooden device is used for hand-breaking. It consists of three boards about 1.5 m long mounted on four legs. Breaking hemp on a hand break is extremely hard work; a good operator can break about 40–50 kg of long fibre per day.

The second method of extracting hemp fibre is by using a mechanical break, followed by turbine scutching. The equipment is not as sophisticated as that used for flax (above), but the principle is the same.

Carefully prepared water-retted hemp fibre is semi-white and lustrous and can be spun into fine yarns.

Sunn fibre

Sunn fibre is retted in a similar way to other stem fibres, but the retting time is usually shorter. In India it can be as short as three to five days in static or slowly moving water.

After retting and drying it is the practice to twist and tie the fibre in small hanks before sending it to market. Fibre which is to be exported is graded after hackling, a process known as 'dressing'. There are numerous grades of sunn fibre based on different areas of production. Fibre is exported in high-density pucca bales weighing 180 kg.

Sunn fibre is white to grey in colour with a shiny lustre and fine texture. It is coarser than jute and not as pliable.

Himalayan or Nilgiri nettle (allo)

The bark is boiled for 3–4 hours in an alkaline solution made from wood ash. This dissolves the bonding between the fibres. The fibres are next rinsed in water and then beaten with a wooden mallet to remove any remaining plant matter and separate the fibres. The resulting fluffy mass of fibres is then rubbed either with a white, mica-rich clay, or with rice husk or corn flour, and hung up in the sun to dry. The application of the white clay in particular seems to have the combined effect of bleaching and lubricating the fibres. The fibres resulting from the above processes are very fine and extremely strong.

Prior to spinning the fibres are opened up, and the fibre bundle is held taut with one end held between the toes and the other end around one arm.

Spinning is done on a lightweight spindle, the whorl carved from wood or bone, the shaft usually from bamboo. A long bundle of fibres 150 cm long and 3–4 cm thick, is tucked around the waist at one end and held under the left arm at the other. Bits of fibre are pulled out using the left hand and teeth, and twisted into the spindle which is held in the right hand.

Nettle yarn is traditionally woven on simple two-shaft or backstrap looms, often used outdoors.

Trials carried out in the 1980s by a number of development agencies demonstrated that by using improved methods of extraction a more refined linen-type of fibre could be produced, for making coarse cloth and belts, ropes and cords.

LEAF FIBRES

Leaf fibres are usually extracted without recourse to retting. In its simplest form the extraction is carried out by first beating the leaf to soften the non-fibrous soft tissue, and then scraping the fibres clean. However, some leaves contain high levels of pectins, and it is then necessary to remove these by retting or chemical extraction.

Sisal

In some parts of the world sisal is still extracted using a very simple scraping device such as that used in Mexico for extracting maguey.

The rate of extraction can be greatly increased using a mechanical device called a raspador. A raspador typically consists of a small covered rotary drum equipped with 8–12 blunt blades and an adjustable breast plate. Decortication is accomplished by hand feeding single leaves halfway into the machine between the rotating drum and breastplate, whilst grasping one end of the leaf. The leaf is then withdrawn from the unit and reversed to clean the second half, while the feeder holds the cleaned fibre portion. Such a decorticator (*Illustration 10*) will produce 135–180 kg of fibre per day when fresh leaves are provided and the waste is taken away by other workers. These

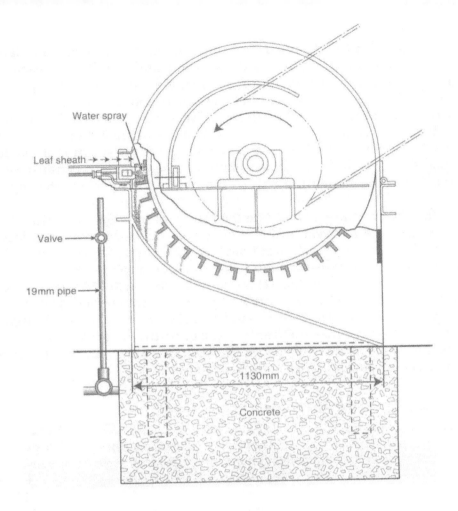

Water spray

Leaf sheath → → →

Valve

19mm pipe

1130mm

Concrete

Illustration 10 Sisal raspador

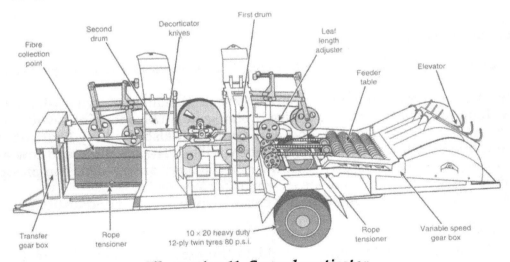

Second drum

Decorticator knives

First drum

Leaf length adjuster

Fibre collection point

Feeder table

Elevator

Transfer gear box

Rope tensioner

10 × 20 heavy duty 12-ply twin tyres 80 p.s.i.

Rope tensioner

Variable speed gear box

Illustration 11 Crane decorticator

machines normally work in the field near where the sisal is grown. Afterwards, the extracted fibre is given a quick wash and hung on a line to dry. This type of extraction is erratic, each leaf being processed separately.

A much higher output is obtained by using a continuous extraction plant in which the leaves are fed automatically into a decorticator and the fibre is discharged at the other end. The cleaning principle of the large automatic 'Krupp' or 'Corona' decorticator is basically the same as that described above, but the machine is much larger and requires around 10 000 gallons of water per hour to lubricate the rapid decorticator action and carry away the waste. The leaves are fed in a layer between link belts or ropes that grip them while they pass through the first drum of the decorticator. They then pass through the 'changeover' section, where the grip is transferred from the leaves to the fibre. The remaining half of the leaf is then decorticated by the second drum.

Large central decorticators necessitate the transport of large quantities of leaves over long distances, and the removal of the waste to the fields. Mobile raspadors overcome these problems but give a lower output and require a large labour input. Recently the Crane double drum mobile decorticator has been developed by an engineer working in East Africa. The Crane (*Illustration 11*) is claimed to be capable of producing fibre from virtually any length of sisal leaf. It is drawn by a 100hp tractor and driven by the power take-off. It operates dry and the fibre has to be washed quickly to avoid discoloration.

It is claimed to be very manoeuvrable and can be driven anywhere a tractor can go on a sisal estate. The output depends on the size and weight of the leaf as well as the power of the tractor (see *Table 2* for comparative figures).

Table 2. *Output for fibre extraction devices*

	Leaves per hour	Weight fibre per day (kg)	Horsepower required
Raspador*	450	30-35	10
Stork type S. 20-12**	30000	5000	160
Crane*		2500	100

*Lock, p.290
**Based on manufacturer's figures.

Note: These figures are approximate. The output will vary with the weight of the leaf and the percentage of fibre in it.

After drying in the sun the fibres, in the case of plantation-grown sisal, are taken to the brushing machines. These are basically similar to a raspador. The revolving blades remove much of the dried pith and epidermis and also the very short fibres. The resultant 'brushing tow 2' is used for mattress filling. The line fibre and tow are then baled for export in bales of 254kg. At the mill the bales are opened. The pre-spinning operations carried out at the mill are shown in *Table 1*.

Well decorticated and brushed sisal is creamy white in colour and lustrous.

Henequen

The leaf is usually taken from the field to the factory by lorry. In Mexico where henequen is grown, the Cordemex central decorticating unit is equipped with sophisticated machinery to ensure full recovery of the fibre from the waste. The decortication plant uses only 20 000 litres of fresh water initially. This water and the leaf juices are recycled to clean the fibre. The solid waste matter which has been removed from the juice is squeezed dry and sent to the nearby dairy unit to feed cattle. The plant

22

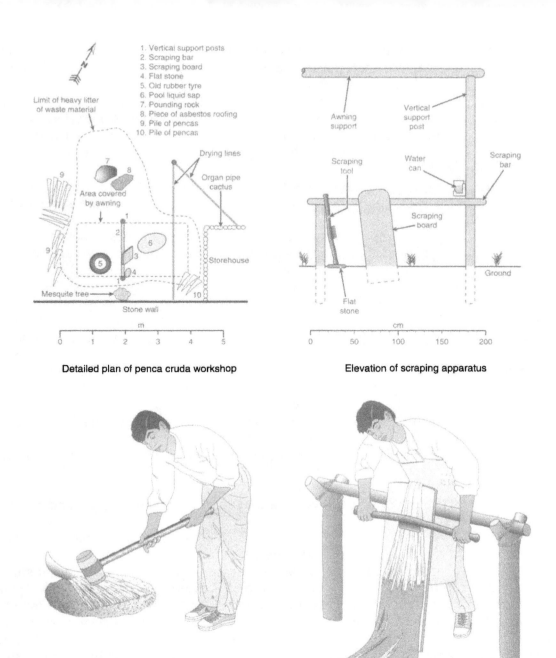

1. Vertical support posts
2. Scraping bar
3. Scraping board
4. Flat stone
5. Old rubber tyre
6. Pool liquid sap
7. Pounding rock
8. Piece of asbestos roofing
9. Pile of pencas
10. Pile of pencas

Detailed plan of penca cruda workshop

Elevation of scraping apparatus

Pounding maguey penca

Scraping maguey penca

Illustration 12 Scraping fibre from maguey leaf

was designed to utilize fully all parts of the henequen leaf, and not only the fibre, which only accounts for approximately 4 per cent of the weight of the leaf.

Henequen is very similar in general appearance to sisal.

Maguey
In Mexico two different methods are used for extracting maguey fibre:

❑   'penca cruda', meaning 'raw leaf', in which the flesh is scraped from the raw leaf
❑   'penca asada', meaning 'cooked leaf', in which the leaf is first heated in a fire
      and left in a damp pit for several days to partially decompose before being scraped.

23

In the 'penca cruda' method, the operator trims the spines off the sides and the thorn from the end of the leaf. He then places the leaf on a large stone and, with a hammer, beats the lower half of the leaf into a flattened pulpy mass.

The operator places this portion on a wooden scraping board and removes the pulp from the fibre using a chopping and scraping action (*Illustration 12*). Periodically the operator pours a little water over the fibre being scraped. Finally, when all the pulp has been removed from the lower half of the leaf, the operator knots the scraped fibre around a twisting rod.

The operator then places the unpounded upper half of the leaf on to the rock surface. The upper half of the leaf is pounded for one or two minutes in much the same manner as the lower half.

The remaining pulp is scraped away in a manner similar to the first. The clean fibre is then lifted off the board, and spread out over a nearby line to dry for several hours. The average leaf requires ten to twenty minutes of labour, and an individual can process twenty-five to thirty-five leaves during a working day of six to eight hours. The yield of dried fibre amounts to between two and three per cent of the original weight of the leaf.

In the 'penca asada' process, two additional items are used; a large hearth where the maguey leaves are heated and 'cooked', and a rotting pit into which the cooked pencas are placed and then covered over for a period of several days.

The hearth is 200 cm in diameter and is raised about 10 to 20 cm above ground level. The rotting pit is a hole dug in the ground 90 cm in diameter and 50 cm deep.

The operator holds each leaf in the fire for about fifteen minutes. The cooked leaves are allowed to cool, and then taken to the horizontal scraping bar where 2 cm is cut off the butt end. The butt end is then beaten with a hammer.

The leaves are laid together in pairs and placed in the rotting pit. When the rotting pit is fully loaded, water is poured in and the top is sealed over with a layer of old leaves, and weighed down with rocks.

After five days the operator removes the leaves from the rotting pit and they are washed and scraped.

Each pair of leaves is then immersed in a tub of water, and the adhering dirt and debris removed by vigorous swirling movements.

After all the leaves in the batch are washed they are placed on top of a clean surface and the actual scraping begins. The scraping process is essentially the same as the 'penca cruda' method; the only two significant differences are the scraping of two (or even three) leaves together, the use of a wooden twisting peg, the complete absence of any leaf pounding, and the occasional use of a knife to trim off bits of charred fibre. Each leaf pair can usually be scraped in seven to ten minutes, much faster than at the 'penca cruda' workshop. As the operator finishes scraping each pair he or she folds the bunch of scraped fibre into a neat pile and places it on a piece of clean plastic. The scraped fibre bundles are then carried over to a nearby line where they are hung up to dry.

Bleaching the fibre In the 'penca asada' process the fibre always has a light yellowish-brown tint after it is scraped and dried. To counteract this, the fibre is bleached in a dilute solution of water, commercial soap and lime juice. The fibre is placed in this solution for a day and then hung to dry outside.

In the Philippines, maguey is traditionally extracted by retting. The bundles of leaves are submerged either in a retting tank or in rivers or creeks. When salt water is used for retting the leaves are first split in half between butt end and tip. When retted in fresh water, the leaves are slit into lengths 1 to 2 cm wide. Retting is faster in salt water, taking 15 days, while in fresh water it takes as long as 20–30 days. Philippine maguey is also extracted using a decorticator. This fibre is lighter in colour than retted fibre.

Again, the further processing of maguey is very similar to sisal.

Pineapple fibre

In the Philippines, the traditional method of extracting the fibre from the leaf makes use of a knife and a piece of porcelain plate. The scraping of the leaf has to be done very carefully to avoid breaking the fibres and the rate of production is very low — as little as 250 g per day. The yield of loose fibre from the fresh leaf is as low as one per cent. In fact, one hundred leaves have to be scraped to obtain only 63 g of fibre. However, no other method is capable of producing the long fine filaments essential for weaving the sheer pina fabric.

The rate of production can be enormously increased if raspadors are used, but the fibre is nothing like so well cleaned as hand-extracted fibre. Quality can however be upgraded by treatment with an alkaline solution, which dissolves the gums adhering to the fibre. The gums can also be removed by retting the leaves. In the course of trials carried out at SITRA (South Indian Textile Research Association) on a raspador designed by KVIC, a production rate of 5 kg in 8 hours was achieved. However, it is reported that an engineering firm in Bombay has developed a power-operated semi-automatic decorticating machine in which the leaves are fed in from one end and collected from the other. It is fitted with a 5hp motor and operates at 1440 rpm. The production capacity is claimed to be 45 kg of pineapple fibre in 8 hours.

Hand extracted fibre The loose fibre scraped from the leaf is composed of roughly one-third long fine filaments (linawan) and two-thirds shorter coarser filaments. The linawan filaments are knotted, by hand, end to end to produce a yarn, and then woven on a handloom to produce very sheer fabrics.

Machine extracted fibre Trials have been carried out mainly at SITRA using a variety of spinning systems, e.g. ring, rotor, semi-worsted, jute, and flax as well as on non-woven systems. The economics are still being investigated.

The quality of the finished fibre depends entirely on the method of extraction. Hand extracted fibre is fine and lustrous, whilst decorticated fibre is coarser and dull.

Abaca

Abaca for textile use is extracted from varieties which yield fine fibre (e.g. maguindanao). It is extracted from the innermost leaf sheath, which yields the whitest fibre.

The harvested pseudostem is separated into its component leaf sheaths. The most desirable fibres are the those which lie just below the outer skin on the rounded surfaces of the leaf sheaths. This outer part of the leaf sheath is removed in the form of strips or 'tuxies'.

Tuxying The tuxies are stripped as follows. A curved knife is used to cut the leaf sheath in such a way that the outer layer is separated from the underlying tissue. This outer layer is held by the end and jerked vigorously to separate it completely. It comes off in the form of ribbons about 5–8 cm wide. The pseudostems are tuxied where they fall in the plantation. Fresh tuxies amount to about 15 per cent of the weight of the leaf sheath. Tuxying reduces the weight of material to be transported. The residue of the stalks, which together with the leaves amount to 85 per cent of the tumbled plant, remains in the plantation to rot, returning humus and mineral nutrients to the soil. A disadvantage of tuxying is that it is laborious and expensive, requiring four times as many person-days per ton of dried recovered fibre as stripping.

Stripping The fibre strands are extracted from the tuxies either by hand or by spindle stripping. The methods for stripping are essentially the same whether carried out by hand or using a spindle. In both cases the tuxies are pulled between a weighted

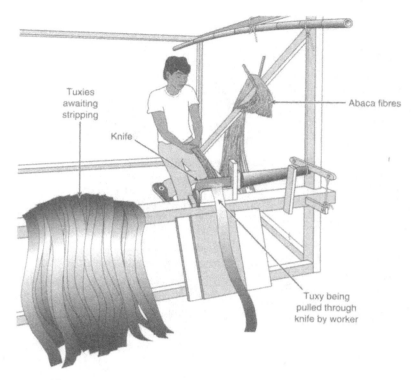

Illustration 13 Hand stripping tuxies from abaca

knife and a block. Tuxy stripping is illustrated in **Illustration 13**. A device called a 'hagotan' constructed mainly from bamboo is used to extract the fibre from the tuxy. It is constructed at very little cost by the farmer from materials he finds locally. Tuxies are fed in topside uppermost.

One end is twisted around a rod about 1 in (2.5 cm) in diameter and 6 in (15 cm) long. The tuxy is then fed between the knife and the wooden block and the end length of about 6 in (15 cm) is twisted round the rod. The knife may be serrated or it may be plain. The plain knife produces better cleaned fibre but requires more effort. The rod and tuxy is gripped firmly, and as the rod is pulled towards the operator the pressure on the tuxy is controlled by the weight of the foot on the treadle. Too much pressure will damage and break the fibre, too little will not remove the skin. It requires experience to produce well-cleaned fibre. After about 20 in (50 cm) of tuxy has been cleaned, the rod is transferred to the cleaned part and the remainder of the tuxy is stripped. Finally, the part of the tuxy twisted around the rod is cleaned, the rod being transferred to the already cleaned fibre.

The spindle stripping machine is factory built and requires power from a diesel engine or electric motor. It is considerably more costly and therefore beyond the reach of many smallholders. The spindle stripping machine is designed to allow the cleaning of six to eight tuxies simultaneously. With hand-stripping, an estate employee could be expected to produce a minimum of 6.5 kg of dry fibre per day, four or five days a week — exceptionally 63 kg a week was obtained. With spindle stripping machines the output could be 23–33 kg per person/day on more days per week. The yield of fibre is affected by many factors including variety, time of harvest and method of fibre extraction but usually 1 to 3 kg are obtained from 100 kg of fresh leaf.

Fibre from the inner leaf sheaths is similar in appearance to sisal, but the fibre is longer. The colour can vary greatly, depending on whether it was extracted from an inner or an outer leaf sheath, from creamy white to brown respectively.

Fique

A very simple device called a 'carrizo' is used for extracting fibre. This consists of two sticks firmly bound in the form of a 'V', fixed to a support or firmly held in the ground. The leaf is split into thin, longitudinal strips. Each one is placed between the two rods, and while it is compressed with one hand it is scraped with the other. The fibre emerges without the fleshy tissue. One worker can extract 10 to 15 pounds (4.5 to 6.8 kg) of fibre in 9 hours' work, but the work is exhausting.

Numerous models of portable decorticator are in use. They all operate on the same principle and vary only in their dimensions, their capacity and the method of feeding. The machine most widely used today has the following characteristics:

The drum is 35 cm in diameter and fitted with 16 knives made of angle iron 28 cm in length. The decorticator is 1 m long, 50 cm wide and weighs 135 kg. Fitted with a 4–6 hp motor its output is 80–120 kg of fibre/day.

After extraction, the fibre is retted for a short time. In small-scale cultivation a watertight wooden tank is used. Plantations of more than 10 000 plants justify the construction of a tank of unburnt brick lined with tiles, since the mucilaginous substances which are given out by the fique strongly attack cement. The interior dimensions of these fermentation tanks are 6.5 ft (2 m) in length, 3.2 ft (1 m) in width and 31 in (80 cm) in height. The capacity is approximately 1.6 cubic metres. In normal conditions a tank of these specifications is perfectly adequate for retting the fibre extracted by a portable decorticator.

The further processing of fique is similar to sisal. Its appearance as raw fibre is likewise similar, although the length of the fibres is shorter.

MANUFACTURE OF YARN

The processes by which any staple raw fibre is converted into yarn are very similar. The following operations are involved:

- ❑ Sliver formation
- ❑ Drafting and doubling
- ❑ Drawing
- ❑ Spinning

The equipment used will vary with the scale of production and the type of fibre: coarse fibres such as sisal are handled on heavier equipment than jute or flax.

Sliver formation is carried out in two distinct ways depending on the mode of presentation of the raw fibre. Cotton comes tightly packed in bales without alignment of fibres whatever, whereas sisal and jute are equally tightly packed but the fibre is approximately aligned and in reasonably discrete parcels or heads.

3. FIBRE QUALITY AND GRADING

The raw quality of fibres — their length, tensile strength, texture and how clean they are — is important when assessing whether they are of export quality and/or their suitability for different end uses.

The methods for deciding on quality vary, depending on fibre type and the region in which it grows. Cotton has internationally recognized grading standards, while other fibres, such as ramie, are subject to variations in grading criteria depending on where they are grown.

The information in this chapter gives a brief outline of the indicators used to assess quality, and the methods for grading raw fibres.

GRADING SYSTEMS

Cotton
Cotton is graded primarily on staple length which varies considerably as can be seen from the examples in *Table 3*.

Table 3. Staple length of different cotton varieties

Variety	Staple length (inches)
Bengals	$1/2$ - $3/8$
Surats	$3/4$ - $11/16$
American Uplands	$3/4$ - $11/16$
Egyptian Uppers	$11/16$ - $1\,1/4$
Egyptian Karnak	$1\,1/4$ - $1\,3/8$
Sea Island	$1\,1/2$ - $1\,3/4$

Coir
Coir yarn is made in hundreds of hamlets and villages along the Kerala coast. This has resulted in a great number of varieties and grades within these varieties. Only a few varieties of yarn are now exported, namely 'Real Superior Anjengo', 'Super Anjengo Magadan', 'Imitation Alapat' and 'Vycome'. Other grades are used locally for woven mat production. There are fewer varieties of white fibre yarn produced in Sri Lanka.

Jute
The current grading system for Bangladesh export jute first separates the two species *Corchorus capsularis* (white jute) and *C. olitorius* (tossa jute). Tossa jute, usually yellowish-red, is frequently softer, stronger and more lustrous than white jute. Fibre from *C. capsularis* varies in colour from pale cream to dull grey or brown and some times dark grey, depending on the amount of iron in the retting water which reacts with the tannin in the stem. However, regardless of the colour, it is usually referred to as 'white jute'.

The highest grades of Bangladesh jute ('Bangla white special' and 'Bangla tossa special') are of the finest texture, very strong and with a high lustre. They are completely free of defects, well hackled and clean cut. The lower grades of jute are weaker and coarser and with bark and specks.

Flax

Linen flax fibre is not generally graded into public marks like other fibres of commerce. Each buyer sets up his own system of grade names or designations. Each of these is represented by a type sample which is changed annually.

Ramie

Ramie is usually sold in the world market in the form of un-degummed ribbons. Different grading systems are employed in the Philippines, Brazil, Japan and China. The Philippines grades, for instance, are as shown in *Table 4*.

Table 4. Ramie grading system

Grade	Length (cm)	Comments
RD-A-Ramie Special	80 or longer	Corona decorticated and washed, good cleaning, straw to creamy colour
RD-1-Ramie good	80 or longer	Corona or raspador decorticated, good cleaning, brownish colour
RD-2-Ramie fair	80 or longer	Washed or unwashed, fair cleaning, light brown
RD-3-Ramie shorts	40–80	Short sound fibre comparable in colour and cleanliness to RD-A, RD-1 and RD-2

Hemp

Hemp is graded mainly on the basis of colour and lustre. The fibre is longer, but coarser and less flexible than flax. No generally recognized trading standards exist. However, in Italy hemp is graded according to a complex system based on a large number of fibre characteristics. The Chilean Association of Hemp producers have set up standards for export grades.

Sunn fibre

Indian sunn fibre is graded under government control according to length, strength, firmness, colour, uniformity and percentage of extraneous matter, called 'refractories'. There are numerous grades of sunn fibre based on the area of production.

Sisal

There are different grading systems for sisal from East Africa and from Brazil. East African sisal is graded according to length and colour as shown in the simplified description in *Table 5*.

Henequen

Henequen is exported in the form of agricultural twine and there is no grading system for raw fibre.

Table 5. Sisal grading system

Grade	Length (cm)/Comments
1	3 ft (0.9 m) or more, creamy white
A	3 ft or more, yellowish, slightly spotted or slightly discoloured
2	2.5–3 ft (0.75–0.9 m) creamy white to cream
3 Long	3 ft or more, minor defects in cleaning and colour allowed
3	2–3 ft (0.6–0.9 m) minor defects in colour and cleaning allowed
UG (under grade)	2 ft or more, defective cleaning and colour allowed
S C W F (short, clean, white fibre)	18 in or more in length
1 tow	Creamy white to cream
2 tow	Darker in colour

Maguey

There are no official grades for Mexican maguey; however, there is a system of principal and secondary grades for Philippine maguey. The principal grades are MR-1 through MR-3, and secondary grades are MR-0 (maguey string), MR-T (maguey tow) and MR-Y (maguey damaged).

Abaca

The fibre is graded before baling based on standards laid down by the Philippine Inspection Bureau. It is usually carried out in warehouses at the point of export. While the fibre is being sorted into grades it is hung over poles, and it is at this stage that the discoloured tips are cut off and consigned to tow. The fibre inspector is concerned with the general colour of the fibre, the presence of dark streaks, the degree of fineness or coarseness and the strength. No instruments are used, although in the event of dispute the colour may be checked and expressed as the 'Becker' value and the tensile strength may be measured. Since the quality is determined partly by the method of extraction, i.e. whether it has been hand- or spindle-stripped and partly by the area of production, namely in Davao or outside (non-Davao), the system is very complex. However, *Table 6* describes the main grades.

Production of abaca started in Ecuador relatively recently and the grading system is far simpler than that employed in the Philippines. This simplification is possible because almost all the fibre produced is spindle-stripped. Grading (from 1 to 5) depends mainly on the colour and the diameter of the fibre strands.

Table 6. Abaca grading system

Grade	Cleaning	Layer of leaf sheath
S2	Excellent	Next to outside
I	Good	Innermost and middle
G	Good	Next to outside
JK	Fair	All except outside
Y2	Damaged during cleaning	All

MECHANICAL PROPERTIES

Table 7. Mechanical properties of plant fibres

Fibre	Length of commercial fibre (mm)	Length of spinnable fibre (mm) (i.e. staple length)	Linear density (Tex)	Tensile strength (kg/sq. mm)	Ext tension at break (%)
Cotton	15–56	15–56			
Coir		20–150	50	30	37
Jute	750–1500	60	1.4–3.0	105	2.7
Kenaf	750–1500	60	1.9–2.2	87	3.5
Roselle	750–1500	60	2.14–3.02	91	3.5
Flax	700–900	50–150 (wet spun)	0.2–2.0	134 183	4.1 3.2
Ramie	800	100–200		129 112	3.9 4.2
Hemp (true)	2500 (long hemp)	150–1500	0.3–2.2	126	4.2
Sunn fibre	750–1000			73	5.5
Himalayan/ Nilgiri nettle (Allo)		8.5–53.6 Mean 32.4 cm	'very strong'		
Sisal	600–1000		28.6–48.6	78	5.0
Henequen			40.2–53.1		
Maguey		300–900	5.0		
Abaca	1000–2000		4.2–44.4	140 85	8.0 7.8
Pineapple	900–1500		1.5–2.3	101 52	4.9 2.4

CHEMICAL COMPOSITION

The chemistry of plant fibres is complex but it is not necessary to go into the subject in depth in order to understand the effect that chemical composition has on the extraction and processing of these fibres. The following is intended only as a very simple account of the chemistry of plant fibres.

All the fibres dealt with in this book contain cellulose; in fact some, such as cotton and ramie, are almost pure cellulose. Chemists recognize several different types of cellulose, namely pure or holocellulose and the hemicelluloses. Holocellulose can resist attack by strong hot alkali which will dissolve the hemicelluloses. Cellulose is a long chain polymer made up of glucose residues joined one to another. The hemicelluloses have shorter chains and are sometimes described as 'encrusting' materials or gums. These gums are present in large amounts in ramie fibre, and in order to obtain spinnable fibre the raw fibre must first be degummed.

The other main compound present in plant fibres is lignin. In spite of exhaustive studies, chemists have not yet succeeded in establishing the structure of this complex compound. Not all plant fibres contain lignin — cotton contains none and ramie and flax contain insignificant amounts, if any. Lignin is present in all woody tissue. It is present in the material which cements plant cells together and when the lignin is dissolved, as happens during the pulping of wood and other paper-making materials, the cells separate from each other to form a suspension which when caught on a sieve forms a paper-like layer.

Lignified fibres turn brown or yellow on exposure to light and this limits their applications in decorative products.

Table 8. Chemical composition (%) of plant fibres

	Cellulose	Hemi-celluloses	Pectin	Lignin	Water solubles	Fat and wax	Moisture
Cotton	82.70	5.70			1.00	0.60	10.00
Jute	64.40	12.00	0.20	11.90	1.10	0.50	10.00
Flax	64.10	16.70	1.80	2.00	3.90	1.50	10.00
Ramie	68.60	13.10	1.90	0.60	5.50	0.30	10.00
Hemp	67.00	16.10	0.80	3.30	2.10	0.70	10.00
Sunn fibre	67.80	16.60	0.30	3.50	1.40	0.40	10.00
Sisal	65.80	12.00	0.80	9.90	1.20	0.30	10.00
Abaca	63.20	19.60	0.50	5.10	1.40	0.20	10.00

Source: Batra / A. J. Turner, *The Structure of Textile Fibres*

Table 9. Chemical composition (%) of Agave and Furcraea fibres

	Alpha cellulose	Ash	Holo-cellulose	Lignin	Pentosans	Solubilities in: Alcohol/benzene	Hot water	0.01 NaOH
Sisal	64.40	0.60	89.40	7.40	1.00	1.00	1.60	14.00
Henequen	67.40	0.80	91.80	5.60	21.70	1.40	0.40	15.00
Maguey	64.30	1.30	87.60	6.60	20.70	1.70	2.80	16.20

Source: Escolana, E. U., Francia, P. C., Semana, J. A., Ynalvez, L. A., 'The proximate chemical composition of some Agave and Furcraea fibres' in *Philippine Lumberman,* Vol. XVI (3) March 1970.

Note: Ash, lignin, alcohol/benzene, hot water and 1% caustic soda solubilities were determined according to TAPPI standards (Technical Association of the Pulp and Paper Industry, *Standards and methods*, New York, n.d.).

Table 10. Chemical composition (%) of pineapple fibre

	Alpha cellulose	Hemicellulose	Lignin	Ash	Alcohol/benzene
Decorticated fibre	79.36	13.07	4.25	2.29	5.73
Retted fibre	87.36	4.58	3.62	0.54	2.27
Degummed fibre	94.21	2.26	2.75	0.37	0.77

Source: Doraiswamy, I. and Chellamani, P., *Pineapple-leaf fibres.* Textile Progress, Vol. 24 No. 1.

Table 11. Chemical composition (%) of coconut fibre

	Water solubles	Pectins	Hemicellulose	Lignin	Cellulose
Young nut	16.00	2.70	0.15	40.50	32.90
Old nut	5.20	3.00	0.25	45.80	43.30

Source: Technical data handbook on the coconut: its products and by-products. Philippine Coconut Authority, 1979.

ASSESSMENT OF FIBRE QUALITY

Freedom from extraneous matter

Cotton and bast and leaf fibres

In addition to visual inspection, the cleanliness of a sample can be determined by measuring the weight loss after boiling it in water and also in an alkaline solution for a fixed period of time.

Fineness

Cotton

Nowadays the fineness of samples of cotton is measured by the Micronaire method in which the resistance presented to a flow of air by a standard size plug of cotton is measured.

Bast and leaf fibres

Fibre fineness is a measure of the diameter of individual fibre strands. It can be judged by opening the structure and examining the strands by hand and eye.

In the laboratory the simplest method of measuring fibre fineness is to determine the linear density by weighing a specified length. The internationally used method of expressing linear density is in terms of Tex. Tex is defined as the weight in grams of one kilometre of fibre.

Tensile strength

Cotton and bast and leaf fibres

Fibre strength can be assessed by snapping a few strands by hand: a qualitative procedure which gives a useful indication to an experienced operator. From the middle portion of the fibre bulk, a bundle of 15–20 clean fibre strands is gripped 5 cm apart between the thumb and forefinger of two hands. Breaking the fibre bundle slowly without jerking gives and indication of fibre strength.

There are several instruments designed to test in the laboratory the breaking load and extension of fibres. These range from fairly basic hand-operated models to very sophisticated automated instruments.

4. SPECIFICATIONS OF SMALL-SCALE PROCESSING MACHINES

There is a vast range of small-scale machinery throughout the world used to process plant fibres; too many to provide a detailed list. The various machines can, however, be divided into the following processing categories. These processes differ mainly according to whether fruit, stem or leaf fibres are being processed.

Cotton

Table 12. Ginning equipment

Ginning	Output in lint per hour
Single roller 40 in	25 kg (55 lb)
Double roller 40 in	35 kg (77 lb)
High capacity roller	150 kg (500 lb)
40-saw	150 kg (500 lb)
80-saw	340 kg (750 lb)
120-saw	500 kg (1100 lb)
Modern 128-saw	over 1000 kg (2200 lb)

Source: Munro, *Cotton.* p.329

Coir (coconut fibre)

Dehusking is normally carried out using a spear firmly fixed in the ground (see *Illustration 6* on page 12). However, mechanical dehuskers have been developed for use in countries where labour costs are high.

The use of a crusher fitted with spiked rollers can greatly reduce the time needed for soaking and retting.

Husk beating to extract the fibre is carried out by hand. No information is available on the mechanical beaters introduced experimentally some years ago.

Table 13. Dehusking equipment

Dehusking	Type of equipment	Output
Manual	Spear fixed in ground	1500–2000 nuts per working day
Mechanical	Jaffa	7000 per day

Table 14. Crushing equipment

Crushing	Type of equipment	Output	Power requirement
Mechanical	Ennor crusher	8000 husks in 8 hours	3.75 kW

Table 15. Husk beating equipment

Husk beating	Type of equipment	Output
Manual	Wooden stick	1000 husks in 10 hours

Table 16. Defibring equipment

Defibring	Type of equipment	Output	Power requirement
Mechanical	Picker drum Ennor Model DF-80	5000 husks in 8 hours	11 kW
Mechanical	Impact mill and sifter Ennor Burster EHD-80	8000–10000 husks in 8 hours	11 kW
Mechanical	Ennor Decorticator DC 80	7000–8000 husks in 8 hours	11 kW

Table 17. Sifting equipment

Sifting	Type of equipment	Output	Power requirement
Mechanical	Revolving screen Ennor RS 80	400–500 kg fibre per 8 hours	0.75 kW
Mechanical	Turbo cleaner Ennor TC -80	400–500 kg fibre per 8 hours	5.6 kW

Defibring of husks can be carried out in two ways.
1. Using picker drums — this produces two types of fibre, one which is longer and coarser and the other shorter and finer.
2. Using an impact mill and sifter. This produces mixed fibre.

Rough cleaning of the fibre is carried out in a revolving screen. Thorough cleaning requires a turbo cleaner.

Flax and jute

There are basically two methods of increasing the rate of stripping from stems. The first employs a very simple manual device such as the TDRI ribboner or the bicycle hub ribboner. The second employs a mechanical rotating drum fitted with blades such as the Alvan Blanch ribboner. We have no data relating to this equipment.

Sisal

Sisal fibre can be extracted using a knife and board, but sometimes a bamboo with a slit is used. A greater output can be achieved using a raspador which uses a rotating drum fitted with blades. The output using a decorticator such as the Crane is much greater than the raspador.

Table 18. Decorticating equipment

Decorticating	Type of equipment	Output (kg of dry fibre)	Power requirement
Manual	Hand scraping		
Mechanical	Raspador	45 in 8 hours	
Mechanical	Crane decorticator	2–3 tons in 8 hours	100 hp

Abaca

Hand-stripping is carried out using a machine constructed from local materials including bamboo, wood, string and a simple knife.

The spindle stripping machine consists of a power source which rotates a spindle to which a knife is attached.

Table 19. Stripping equipment

Stripping	Type of equipment	Output (dry fibre)
Manual	Hagotan	23 kg in 8 hours
Mechanical	Spindle stripper	65 kg in 8 hours

5. PLANNING FOR PRODUCTION

The circumstances for processing any plant fibre are likely to vary widely. Most plants are normally processed locally, close to where they are grown, before being transferred as fibre to other factories, locally or abroad, and then manufactured into end-products. Some of the essential issues which must be considered before setting up a plant fibre processing unit include:

- ❏ Source of plants (tropical or temperate countries)
- ❏ Supply of plants (constant cultivated, or irregular grown wild)
- ❏ Human resources (local or imported skills in machine and building construction)
- ❏ Social acceptability (demand for the product from within the community or from elsewhere?)
- ❏ Consultation (must be locally acceptable, suitable and practical)
- ❏ Environmental impact (are local water supplies or land likely to be polluted? Waste disposal?)
- ❏ Employment potential (need for expansion of existing employment, or creating new jobs?)
- ❏ Income (creating or adding to personal, family and local economy)

The previous chapters show the wide variety of plants that are grown for the purpose of extracting their fibres for textile end uses. There are many well-established small-scale fibre processing units all over the world. Perhaps, therefore, the intention might be to expand, adapt or improve existing production. It may, on the other hand, be the intention to establish a processing unit in a new situation, where there is little previous experience.

Sometimes plant fibres are grown specifically for large commercial end uses and their production is only viable on a large scale. On the other hand, some indigenous plants can be processed on a small scale. Whatever the situation, careful planning is important, and attention to the basic considerations listed above should help to achieve a successful result.

MARKET CONSIDERATIONS

It is essential to have a solid understanding of the needs and markets for the product before manufacture can be planned. The first, important, step is to carry out a market survey aiming to acquire the following information.

- ❏ What are the potential markets for the fibre you are planning to process?
- ❏ What is the current demand for the type of fibre to be processed?
- ❏ What is the competition from other producers and alternative fibres?
- ❏ Will the supply meet the demand and is there room for future growth?
- ❏ What are the distribution channels, from processor to fibre user?
- ❏ What are the existing prices for the processed fibre and are there any invisible costs such as duties or tariffs?

IMPORTANCE OF PLANNING

The market survey is the most important first step to be taken and the results should provide a clear indication either to go ahead or think again. Before making any decisions about the scale and type of equipment needed in a new situation or updating an already well-established enterprise, the facts need further careful consideration.

If the decision at this stage is to go ahead, many further questions will arise, requiring positive answers as part of the planning process, and before taking the next step in actually setting up a plant fibre processing enterprise.

❑ What is the size of the planned enterprise and will it have the most appropriate equipment for the job?
❑ Is there enough capacity for growth on the chosen premises ?
❑ Is there sufficient justification or need for the level of expenditure planned?
❑ Is there sufficient cash available to meet the initial purchasing costs, including spare parts and ancillary equipment, wages, transport and day-to-day running expenditure?
❑ Is it possible to find reliable advice and guidance regarding sourcing loans or credit?
❑ Are the fibre-producing plants cultivated locally, and in what quantities?
❑ Will the plant fibre processing enterprise be dependent solely on commercial growers?
❑ Will the raw material be irregularly supplied by small growers or regularly collected by your operation?
❑ Will you have the necessary experience to set up all aspects of the business. If not, where will you find this?
❑ Will the processing equipment be easy to use? Can any necessary training be obtained locally, or at a national training centre?
❑ Do appropriate local skills exist, or could they be learned through training, for the maintenance and servicing of equipment?
❑ Where will the plant fibre processing unit be sited and will it be within easy reach of the market place?
❑ Will the enterprise have a positive or detrimental effect on the local environment, and what steps can you take to offset or prevent any harmful effects?
❑ Are you able to apply the strict legal requirements for safety and hygiene codes of practice?

FINANCIAL FEASIBILITY

Once the operation has been planned in some detail, it is useful to undertake a trial costing for production to provide a rough idea of its commercial viability. The cost of production generally has two components: fixed overheads or indirect costs, and variable or direct costs.

Overheads or indirect costs

❑ Interest on the cost of any stock of raw materials
❑ Premises (rent or purchase and insurance)

- Heat/light/water/power
- Telephone/fax/postage/stationery
- Depreciation of equipment and interest on purchase loans
- Consumable materials
- Administration costs
- Marketing costs

Direct or variable costs

- Raw materials (harvested plants ready for processing)
- Waste (re-saleable as fuel, fodder or fertilizer?)
- Transport (to and from the processing unit)
- Wages (including any contribution to a welfare fund and any incentive wages)

Break-even analysis

To estimate the production output needed to compete in the market, first of all assess its financial feasibility. A break-even analysis considers the following questions:

- At what level of production will the product be able to cover its direct and indirect costs?
- What is the minimum price per kg or tonne needed for the project to be able to cover its fixed and variable costs?
- What is the minimum price needed for the project to break even?
- What happens if financial assumptions or actual costs or prices are changed?
- What are the best, worst, and probable scenarios for the project?

The estimated production output needed to break even should be compared with the capacity of the equipment and with the size of the expected market. To check the viability of the planned unit, the cost of production should also be compared with prices in the market.

ECONOMICS

There is a direct correlation between the quantity of processed fibre and profit. However, production cannot be increased indiscriminately without compromising quality. It is important to recognize that production capacity will be limited by the infrastructure, the plants that will grow locally, and the demand for processed fibre. Production should be planned to make the unit economically viable into the future.

CONCLUSION

Guidance should always be sought from local manufacturing and business expertise, if it is available, making sure you have as much information as possible before proceeding to set a processing operation.

Do not forget that the time taken in planning and in consultation at every stage will be rewarded with a successful business.

6. EQUIPMENT SUPPLIERS

Alvan Blanch Development Company Ltd Chelworth Malmesbury Wilts SN16 9SG UK	Stem fibre ribboner
Chainbrace Ltd The Manor Whitestaunton Chard Somerset TA20 3DL UK	Mobile sisal decorticator
Vicente F. Duhaylungsod Machine Shop 106R Castillo Street Agado Davao City Philippines	Abaca spindle stripping machine and ramie decorticator
George C. Dy, Jr. 7 Florencia Street San Francisco del Monte Quezon City Philippines	Small-scale fibre extraction machine
Dynamic Machine Shop R. Castillo Street Agdao Davao City Philippines	Abaca spindle stripping machine and ramie decorticator
Ennor Industries Private Ltd 37 Graemes Road Thousand Lights Madras 600 006 India	Coconut fibre extraction and processing machinery
Godwell Engineering Products Avanashi Road Civil Aerodrome Post Coimbatore 641 014 India	Coconut fibre extraction and processing machinery

Hilanderias del Fonce Calle 35 No 17-56 Bucaramanga Colombia	Mobile raspadors
Khadi and Village Industries Commission Irla Vile Parle Bombay 400056 India	Wide range of small-scale fibre processing equipment
Luna's Machine Shop Lasang Davao City Philippines	Abaca spindle stripping machine and ramie decorticator
Mackie International Ltd 385 Springfield Road Belfast BT12 7DG N. Ireland	Fibre processing machinery
SDL International Ltd PO 162 Crown Royal Shawcross Street Stockport SK1 3JW UK	Test equipment
Wm. R. Stewart & Sons [Hacklemakers]Ltd Marine Parade Dundee Scotland DD1 3JD UK	Fibre extraction equipment
Consumables A. J. Dickenson Ltd Neptune Chemical Works London SE6 5JF UK	Batching emulsions

7. SOURCES OF FURTHER INFORMATION

RESEARCH ORGANIZATIONS

International Jute Organization
95A, Road No 4
Banani
PO Box No 6073
Gulshan
Dhaka, 1213
Bangladesh

Instituto Agronomico de Campinas
Secao de Plantas Fibrosas
Caixa Postal 28
13001 9 970 Campinas
SP - Brazil

Instituto de Pesquisas Tecnologicas S A
DPFT -CETEC
R. Prof Almeida Prado, 532
Caixa Postal 101064 - 970 Sao Paulo
SP - Brazil

Instituto Nacional de Pesquisas de Amazonia INPA
Cx. Postal 478
69.011 Manaus, AM
SP - Brazil

EMGOPA, Caixa Postal 49
CEP 74130
Goiania, GO.
Brazil

West Canton Experimental Station of South China
Academy of Tropical Crops
Zhanjiang
Guangdong
China

Guangxi Institute of Tropical Crops
Nanning
Guanx
China

Institute of Ramie Research
Daxian Region
Sichuan Province
China

Institute of Bast Fibre Crops
Chinese Academy of Agricultural Sciences
Yuanjian County
Hunan 413100
China

Institute of Agricultural Science
Wanrong County
Shanxi
China

Institute of Fibre Crops
Beijing
China

Fujian Academy of Agricultural Sciences
Fuzhou
Fujian
China

GREF
Ministere de l'Agriculture et de la Foret
Paris
France

Inst. Pflanzenbau und Pflanzenzuchtung
Bundesforschungsanstalt Braunschweig-Volkenrode
3300 Braunschweig
Germany

Appropriate Technology Development Association
Lucknow
India

Indian Jute Industries Research Association
Tartola Road
Calcutta
India

Jute Technological Research Laboratory
Calcutta
India

South India Textile Research Association
PB No. 3205
Coimbatore Airport PO
Coimbatore 641 014
South India

Regional Research Laboratory, CSIR
Bhopal 462026
Madhya Pradesh
India

Israel Fiber Institute
5 Emek Refaim Street
PO Box 8001
Jerusalem
Israel

Agricultural Research Department
Agrotechnical Research Institute [ATO-DLO]
Haagsteeg 6
PO Box 17
NL-6700 AA Wageningen
Netherlands

Center for Agrobiological Research
6700 AA Wageningen
Netherlands

Fiber Industry Development Authority
Philfinance Building
Benavidez Street
Legaspi Village
Makati
Metro Manila
Philippines

Philippine Textile Research Institute
General Santos Avenue
Bicutan
Paranaque
Metro Manila
Philippines

University of the Philippines at Los Banos
Laguna
Philippines

Visayas State College of Agriculture
Baybay
Leyte
Philippines

Philippine Coconut Authority
Diliman
PO Box 386
Quezon City
Philippines

Coconut Information Centre
Coconut Research Institute
Lunuwila
Sri Lanka

AFRC
Silsoe Research Institute
Wrest Park
Silsoe
Bedford MK45 4HS
UK

Silsoe College
Cranfield University
Silsoe
Bedford MK45 4DT
UK

BTTG
Shirley Towers
856 Wilmslow Road
Didsbury
Manchester M20 8RX
UK

Intermediate Technology
Myson House
Railway Terrace
Rugby CV21 3HT
UK

Natural Fibres Organisation
The Town House
The Square
Bishop's Waltham
Hampshire SO32 1AF
UK

Natural Resources Institute [NRI]
Central Avenue
Chatham Maritime
Chatham
Kent ME4 4TB
UK

Department of Textile Industries
University of Leeds
Leeds LS2 9JT
UK

Economic and Conservation Section
Royal Botanic Gardens Kew
Richmond
Surrey TW9 3AB
UK

Department of Agricultural Botany
The Queen's University of Belfast
Belfast BT9 5PX
UK

Department of Biological and Biomedical Sciences
University of Ulster
Coleraine
Co. Derry BT52 1SA
UK

Plant Pathology Research Division
Department of Agriculture for Northern Ireland
Newforge Lane
Belfast BT9 5PX
UK

School of Materials Science
University of Bath
Bath
UK

OTHER USEFUL SOURCES OF INFORMATION

Asian and Pacific Coconut Community
Wisma Bakri
J. H. R. Rasuna Said
Kuningan
Jakarta
Indonesia

Coir Board
M G Road
Cochin 682016
India

Food and Agriculture Organisation
Via delle Terme di Caracalla
00100 Rome
Italy

International Trade Centre
Palais des Nations
1211 Geneva 10
Switzerland

Kenya Sisal Board
Old Mutual Building
Kimathi Street
PO Box 41179
Nairobi
Kenya

Tanzania Sisal Authority
PO Box 277
Tanga
Tanzania

UNIDO
PO Box 300
A - 1014 Vienna
Austria

PUBLISHED SOURCES OF INFORMATION

Acland, J. D., *East African Crops*. Longman, 1971.

Amalraj, V. A., *Cultivated sedges of South India for mat weaving industry*. J. Econ. Tax. Bot., Vol. 14 no. 3, pp. 629–631, 1990.

Andreas, B., *The economics of tropical agriculture*. English edition edited by Jean Kestner. Commonwealth Agricultural Bureau, 1980.

Atkinson, R. R., *Jute-fibre to yarn*. Temple Press, London, 212 pp., 1964.

Batra, S. K., 'Other long vegetable fibers in fibre chemistry', in *Handbook of Fiber Science and Technology*: Volume IV. Lewin, M. and Pearce, E. (eds.), Dekker Inc., New York and Basel, 1983.

Canning, A. J. and Green, J. H. S., 'Nepalese 'allo' (Girardinia): improved processing and dyeing' in *Tropical Science*, Vol. 26, pp.79–82, 1986.

Davis, J. B., Kay D. E. and Clark V., *Plants tolerant of arid, or semi-arid, conditions with non-food constituents of potential use*. Tropical Development and Research Institute, 172 pp., 1983.

Dempsey, J. M., *Fibre Crops*. Gainesville: University of Florida Presses., 457 pp., 1975.

Dodge, C. R., *A Descriptive Catalogue of Useful Fiber Plants of the World*. US Department of Agriculture, 1987.

Doriaswamy, I. and Chellamani, P., 'Pineapple-leaf fibres', in *Textile Progress* Vol. 24 No. 1. Textile Institute, Manchester, 1993.

Himmelfarb, D., *The technology of cordage fibres and rope*. Leonard Hill [Books] Ltd., 1957.

Jarman C. G., Canning A. J. and Mykoluk, S., 'Cultivation, extraction and processing of ramie fibre: a review', in *Tropical Science*, Vol. 20 (2), pp.91–116, 1978.

Jarman C. G. et al., 'Banana fibre: a review of its properties and small scale extraction and processing', in *Tropical Science*, Vol. 19 (4), 1977.

Jarman C. G. and Robbins S. R. J., *An industrial profile of coconut fibre extraction and processing*. TDRI Report G189, Oct. 1986.

Jarman, C. G., *The retting of jute*. FAO, Rome Agricultural Services Bulletin No. 60, 1985.

Kirby R. H., *Vegetable Fibres*. Leonard Hill, London, 464 pp., 1963.

Lock, G. W., *Sisal*. Longman, London and Harlow, 2nd ed. 1969.

Morton, W. E. and Hearle, J. W. S., *Physical Properties of Textile Fibres*. Textile Institute, 3rd ed. 1993.

Munro, J. M., *Cotton*. Longman, 2nd ed. 1987.

Palmer, E. R., 'The use of abaca for pulp and paper making', in *Tropical Science*, Vol. 24(1), pp.1–16, 1982.

Parsons, J .R. and Parsons, M. H., *Maguey cultivation in Highland Central Mexico*. Museum of Anthropology, Michigan, Ann Arbor, 1990.

Picton, J. and Mack, J., *African Textiles*. British Museums Publications Ltd., London, 1979.

Perez, J. A. *El fique*. Compania de Empaques, S A Medellin, Colombia, 2nd ed. 1974.

Perry, D. R. and Farnfield, C. A., *Identification of Textile Materials*. Textile Institute, 7th ed. 1985.

Purseglove, J. W., *Tropical Crops: Dicotyledons*. Longman, 1968.

Ramaswamy S. S. and Muthukuraswamy, M., *High production decorticating machine for extraction of fibre from pineapple leaves*. The Institution of Engineers (India), Platinum Jubilee Celebrations, Seminar: Technology for a better tomorrow, Calcutta, Dec. 15–29, 1994.

Royal Botanic Gardens, Kew. *The Thread of Life: the story of our use of cellulose*, ISBN 0 946743 23 0, n.d.

Sinha, Frances, *Allo processing in Nepal*. Intermediate Technology Development Group, 40 pp., 1989.

Stout, H. P., 'Fibre and yarn quality in jute spinning', in *Manual of Textile Technology*. The Textile Institute, Manchester, ISBN 1 870812 09 3, 1988.

Textile Terms and Definitions. The Textile Institute, Manchester.

Wilson, P., *Sisal: a study of the plant and its leaf fibre*. Hard Fibres Research series No. 8, FAO Rome, 1971.

APPENDIX 1

It would be quite impossible to list all the plants which yield a useful fibre — over 350 have been listed in East Africa alone. However, an attempt has been made to include most of the fibre plants mentioned in the literature as well as those yielding the well-known fibres of commerce.

Distribution and uses of fibre-yielding plants

Common name	Botanical name	Distribution	Main uses of fibrous parts	Process in similar way to:	Other uses
Abaca	*Musa textilis* Née Musaceae	Philippines, Ecuador	Ropes, cloth for handicrafts	Sisal	Fibre used for speciality papers
Akund	*Calotropis procera* (Ait.) Ait. f. Asclepiadaceae	India, Africa, Australia, Brazil	Seed fibres used for upholstery stuffing, bast fibres used for rope	Kapok (seed fibres)	Source of latex
Allo (Nilgiri or Himalayan nettle)	*Girardinia diversifolia* (Link) I. Friis Urticaceae	Himalayas	Fishing nets, sacks bags, nets, hand-woven cloth	Ramie	
Aloe	see Mauritius hemp				
Bahama hemp	see Sisal				
Bamboo reed	*Arundo donax* L. Gramineae	Native to Mediterranean but now widespread in sub-tropics	Basketry, matting, musical reeds, industrial cellulose		Rhizomes used medicinally Cane for fishing rods
Banana fibre	*Musa* spp. and *Ensete* spp. esp. *E. ventricosum* (Welw.) Cheesman Musaceae	South India to Japan and Samoa, South and West Africa, Brazil	Rope, cloth	Abaca	Fruit, starch foodstuff and boiled vegetable
Baobab or Monkey bread tree	*Adansonia digitata* L. Bombacaceae	Africa	Ropes and sacking (inner bark)		Outer bark used as cloth
Basswood or linden or lime	*Tilia* spp. Tiliaceae	N. temperate regions	Fabrics, baskets (bast)		Timber
Bimlipatam jute	see Kenaf				
Bombay hemp	see Sunn fibre				
Bowstring hemp	*Sansevieria* spp. Dracaenaceae	Africa, tropical Asia, Mexico and South America	Cord, rope, sails, paper	Sisal	
Buri palm	*Corypha utan* Lam. Palmae	Philippines	Fibre from petioles (buntal) used in handicrafts		Strips of epidermis yield raffia
Buriti palm	*Mauritia flexuosa* L. f. Palmae	North and South America, Trinidad	Rope, weaving	Thread-like fibre stripped from outer skin of young leaves	Shoots yield 'hearts of palm'. Edible fruit yields oil rich in vitamins. Pith of trunk yields a sago-like starch. Buriti wood used in rafts. Thatch

Common name	Botanical name	Distribution	Main uses of fibrous parts	Process in similar way to:	Other uses
Cabuya	*Furcraea cabuya* Trel. Agavaceae	C. America	Ropes, hammocks	Sisal	
Cantala	*Agave cantala* Roxb. ex Salm-Dyck Agavaceae	Philippines, Indonesia	Fibre softer and finer but weaker than sisal	Sisal	
Canton fibre	*Musa* hybrid Musaceae	Philippines	Used as abaca (lower quality)	Abaca	
Canton linen	see Ramie				
Caranday palm	*Copernicia alba* Morong or *Trithrinax* spp. Palmae	North and South America	Fibre used for ropes, hammocks and hats	Jute	Wax extracted from the leaves
Carnauba wax palm	*Copernicia prunifera* (Miller) H. Moore Palmae	Brazil	Fibre used in construction of domestic articles and for stuffing mattresses		Wax extracted from the leaves. Yields edible fruit, oil from the seed, pith flour, sweet sap juice and palm hearts.
Caroa	*Neoglaziovia variegata* (Arruda) Mez Bromeliaceae	Brazil	Nets	Jute	Paper, silk-like fabric
Cat-tail	*Typha* spp. Typhaceae	Temperate and tropics	Paper, mats, chair seating		
Ceylon piassava	*Caryota urens* L. Palmae	Indo-Malaysia	Brushes	Jute	
China grass	see Ramie				
China jute/hemp	*Abutilon theophrasti* Medik. Malvaceae	China, India, South and Central America	Sacks and hessian	Jute	
Coir (Coconut fibre)	*Cocos nucifera* L. Palmae	Tropics	Twine, rope, mattress fibre, brushes		
Congo jute	see Urena				
Cotton	*Gossypium* spp. including many hybrids, e.g. Short fibre: *G. herbaceum* Long fibre: *G. barbadense* Malvaceae	USA, former USSR, China, S. America, Africa, India, West Indies, Egypt, Sudan	Textiles, clothing		Oil makes up 20% of the seed. Cotton seed oil is used in margarine and soap. Cotton seed cake, the residue after milling, is a valuable stock feed. Seed oil identified as male contraceptive!
Cowpea	*Vigna unguiculata* (L.) Walp. Leguminosae	Tropics	Peduncle yields fibre		Pods and seed used as food. Valued as cover or green manure crop
Crin végétal	*Chaemerops humilis* L. Palmae	N. Africa Mediterranean	Upholstery, ropes and tent fabrics		
Date palm fibre	*Phoenix dactylifera* L. Palmae	Tropics	Ribbons from the leaves are used for weaving very fine mats		Fruit
Deccan hemp	see Kenaf				
Dhaincha	*Sesbania bispinosa* (Jacq.) W. Wight Leguminosae	India, tropics	Paper pulp, sails and fishing nets		Rural fuel. Important leguminous green manure.

Distribution and uses of fibre-yielding plants - continued

Common name	Botanical name	Distribution	Main uses of fibrous parts	Process in similar way to:	Other uses
Dog bane	*Apocynum cannabinum* L. Apocynaceae	N. America	Ropes and sails. Bark used for cordage fibre. Whole stem for textile fibre.		Rhizome used as cardiac stimulant and diuretic/emetic
Doum or Dom palm	*Hyphaene* spp. Palmae	Africa and Middle East	Fibre from leaflets used for sack making		Vegetable ivory from kernel
Esparto grass	*Stipa tenacissima* L. Gramineae	Mediterranean	Ropes, sails, mats		Paper making. Wax used in polishes
Fique	*Furcraea* spp. Agavaceae	Tropical America	Cordage, pulp and paper, carpets and fibre crafts		
Flax	*Linum usitatissimum* L. Linaceae	Europe, Russia	Clothing fabrics, household linens, canvas, ropes, sacks		Paper, linseed oil
Hemp	*Cannabis sativa* L. Cannabaceae	Temperate areas and tropics	Twine, rope, tarpaulins, bed sheets, table cloths		Alkaloid resin (cannabis) produced from some varieties
Henequen	*Agave fourcroydes* Lem. Agavaceae	Mexico	Agricultural twine, cordage, upholstery padding, floor covering	Sisal	
Himalayan giant nettle	See Allo				
Hybrid 11648	Hybrid of *Agave angustifolia* x *A. amaniensis* Agavaceae	East Africa and S. America	Twine, rope, paper, sacking	Sisal	
Indian hemp	see Sunn fibre				
Ita palm	see Buriti palm				
Ixtle or Ixtli	*Agave* spp. and *Yucca carnerosana* (Trel.) McKelvey Agavaceae	Mexico, Texas	Fibre from the inner leaves of the rosette used for brush filling		Soap substitute, paper making
Java jute	see Kenaf				
Jubblepore hemp	see Sunn fibre				
Jute	*Corchorus capsularis* L. White or China jute *Corchorus olitorius* L. Brown or tossa jute Tiliaceae	Tropics	Sacks and hessian, carpet backing, twine		Paper. Stems ('stick') used for fencing and fuel
Kapok	*Ceiba pentandra* (L.) Gaertner Bombacaceae	Tropical America, Africa, India, Java	Stuffing for cushions, mattresses and life jackets. May be mixed with cotton and wool and spun into a yarn.		
Kenaf	*Hibiscus cannabinus* L. Malvaceae	Tropical Africa, India, SE Europe	Twine, sacks, hessian	Jute	Seed oil for fuel
Ketaki	*Pandanus tectorius* Sol. ex Parkinson Pandanaceae	Coastal areas in Europe, Asia, Africa	Mat making, baskets (Most *Pandanus* spp. are used for fibres)		

Common name	Botanical name	Distribution	Main uses of fibrous parts	Process in similar way to:	Other uses
Kitul or Kittool	*Caryota urens* L. Palmae	Indo-Malaysia	Sewing thread, fishing nets and lines, brushes		Sago and palm sugar
Linseed	see Flax				
Madras hemp	see Sunn fibre				
Maguey	*Agave* spp. Agavaceae	Mexico, Philippines	Cordage, pulp and paper, carpets and fibre crafts	Sisal	Leaf pulp yields 'pulque' from which alcoholic drink is produced by fermentation
Malva	*Sida* spp. Malvaceae	Australia, Brazil, India, Argentina, W. Africa	Resembles jute; less strong but softer		
Manila hemp	see Abaca				
Mauritius hemp	*Furcraea foetida* (L.) Haw. Agavaceae	Central and S. America, Mauritius and Reunion	Twine, cordage, sacking	Sisal	Used as fire break in Sri Lanka
Mexican fibre	*Agave lechuguilla* Agavaceae	N. Mexico, Southern USA	See Maguey		
Milkweed	*Asclepias* spp. Asclepiadaceae	N. and C. America	Stuffing (seed and bast fibres)		Medicines, latex (rubber)
Napunti fibre	*Clappertonia ficifolia* (Willd.) Decne. Tilliaceae	W. Africa	Sacks, jute substitute		
New Zealand hemp	*Phormium tenax* Forster & Forster f. Phormiaceae	New Zealand	Tachiko (Maori) weaving, cordage, nets		
Nilgiri nettle	see Allo				
Palmetto palms	*Sabal* spp. Palmae	N. and S. America	Braiding and weaving		
Palmyra fibre or bassine	*Borassus flabellifer* L.	India to Burma	Tender leaves used for mats and basket ware. Material for weaving and thatching.		Food and beverages
Panama hat palm or toquillo	*Carludovica palmata* Ruiz & Pavon Palmae	American tropics	Panama hats (young leaves), brushes, mats, baskets (older leaves)		
Pandanus	see Ketaki				
Paper mulberry	*Broussonetia papyrifera* (L.) Vent Moraceae	Asia, N. America	Cloth (tapa)		
Papyrus	*Cyperus papyrus* L. Cyperaceae	C. Africa	Paper (leaf pith)		Sandals, ropes

Common name	Botanical name	Distribution	Main uses of fibrous parts	Process in similar way to:	Other uses
Piassava	*Raphia hookeri* G. Mann & H. Wendl *Attalea funifera* C. Martins ex Sprengel Palmae	West Africa Brazil	Stiff brushes	Jute	
Pineapple fibre	*Ananas comosus* (L.) Merr. Bromeliaceae	Tropics	Traditional clothing, fine embroidery		
Pita fibre or silkgrass	*Chevaliera magdalenae* (André) André Bromeliaceae	N., C. and S. America	Rope, twine, thread for sewing leather	Jute	
Punga	*Triumfetta rhomboidea* Jacq. Tilliaceae	Brazil, W. Africa	Twine, rope	Jute	
Ramma	see Kenaf				
Ramie	*Boehmeria nivea* (L.) Gandich. Urticaceae	Tropical Asia, Southern USA, Mediterranean	Clothing fabrics, rope, netting, upholstery fabric, canvas, bed sheets, gas mantles		
Raphia or Raffia	*Raphia* spp. Palmae	Tropical America, Africa, Madagascar	Gardening ties, ropes, fishing tackle, cloth, upholstery, decorative household items		
Rattan	*Calamus* spp. *Oncocalamus* spp. *Eremospatha* spp. and others Palmae	Tropics	Bowstrings, shields and bindings, baskets		
Roselle	*Hibiscus sabdariffa* L. Malvaceae	Tropics	Rope	Jute	Foodstuffs
Screw pines	see Ketaki and Panama hat palm				
Sea grass	*Zostera marina* L. Zosteraceae	Europe	Matting, padding		
Sisal or Bahama hemp	*Agave sisalana* Perrine Agavaceae	East Africa, Brazil, C. America	Agricultural twine, ropes, reinforcement of cement sheets, heavy sacking, floor coverings		Planted for erosion control, leaf pulp used as livestock feed, paper
Spanish or Florida moss	*Tillandsia usneoides* L. Bromeliaceae	Gulf Coast of N. and S. America	Stuffing (upholstery)		
Sunn fibre	*Crotalaria juncea* L. Leguminosae	India, Sri Lanka	Ropes, nets, twine, cordage, paper, canvas	Hemp	Widely grown as green manure
Urena	*Urena lobata* L. Malvaceae	W. Africa, S.E. Asia, S. America, Caribbean	Jute substitute	Jute	
Vegetable horse hair	see Crin végétal				
Yucca	*Yucca baccata* Torr. Agavaceae	S. USA, Mexico	Basketry, mats and tying purposes		

APPENDIX 2

GLOSSARY

Angiosperm. The plant kingdom comprises plants which have flowers and those such as fungi and algae which do not. The flowering plants are sub-divided into the gymnosperms, in which the seed is exposed, and the angiosperms, in which the seed is enclosed within an ovary.

Bale breaker. Jute is baled under very high pressure. Before it can be processed it has to be broken down again into bundles of fibres which can be handled by the carding machine. This is done by passing it through a series of heavy inter-meshing toothed rollers, known as a bale breaker.

Bast fibres. Fibre obtained from between the inner and outer layer of the stems or stalks of many plants such as: allo, flax, hemp, hop, jute, kenef, nettle, ramie, roselle, sunn hemp, urena. They are strong, long fibres and can be used in the production of ropes, string, gunny, hessian, sacking and fishing nets.

Batching. (1) The process of selecting and mixing different grades of (for example) jute to give yarn of the required quality at the lowest cost. (2) Adding batching emulsion to jute or sisal to soften the fibre and thereby ease its passage through the spinning line.

Batching oil. Used in the form of an emulsion with water to make batching emulsion.

Butting. To level the root ends of flax straw, ramie, or similar bast fibre producing plants, by vibrating it upright on a flat surface, either by hand or mechanically.

Carding. The cleaning and conversion of raw opened fibre from the bale into a mat of parallel fibres. Carding is often carried out in several stages. In the case of jute the first carding stage is called a 'breaker card'. This breaks the continuous mesh of 'entities'. These 'entities' are akin to the single fibres of cotton or wool. It is important to distinguish between these spinnable fibres which vary in length from a few millimetres in length up to 500 mm and the so-called 'ultimate fibres' which result from the chemical breakdown of the 'entities' which average about 2.5 mm long and are clearly unspinnable on any currently available spinning system.

Cellulose. The basic material from which plant tissues are formed. Cellulose consists chemically of glucose molecules joined end to end in long chains. The glucose is formed by the reaction of carbon dioxide and water in the presence of chlorophyll which acts as a catalyst.

Combing. Before fibres can be spun they have to be aligned parallel to one another. This operation also cleans the fibres.

Dicotyledonous. The cotyledon is the first formed leaf of a plant. *Angiosperms* (see above) are divided into those plants which have two seed leaves (the dicotyledons) and those with one seed leaf (the monocotyledons). Dicotyledonous plants have net-veined leaves.

Doubling. Combining two slivers in order to obtain a sliver with fewer weak spots. Doubling can be repeated several times.

Drafting. This is the name given to the process which follows the process of roving (q.v.). During drafting the fibres are extended and drawn over each other until the desired thickness is achieved. A twist is then inserted to form a yarn.

Ginning. The removal of the seed hairs (lint) from the cotton seed. The remaining hairs are called 'linters'. They are too short to spin, but are used to make paper and as a source of cellulose for converting into regenerated cellulose ('rayon', 'viscose').

Goods machine. Used in the processing of hard fibres. The first Goods machine acts as a breaker in which the fibre is combed, drafted and converted into a continuous sliver (q.v.). The second Goods machine is also called a 'spreader'. The fibre is given a further combing, doubled and drafted.

Hackling. Is a process similar to combing, the main object being to remove the shorter fibres and adhering non-fibrous matter.

Hard fibre. The common fibres extracted from plants are designated by fibre merchants as 'hard' or 'soft'. In general the hard fibres are extracted from the leaves of monocotyledonous (q.v.) plants whilst the soft fibres are derived from the stems of dicotyledonous (q.v.) plants. The stem fibres are finer than fibres from the leaves of monocotyledonous plants.

Head. One of a number of bunches of raw jute forming a bale. The heads are each given a twist and folded over before being made into a bale.

Lignin. A complex molecule present in wood and many plant fibres. Its structure is still being studied. Cotton and flax contain very little lignin, jute and sisal contain around 15 per cent whilst coir contains as much as 40 per cent. Lignified fibres turn yellow/brown on exposure to light and dyes fade quickly.

Monocotyledonous. Having only one seed leaf. Monocotyledonous plants tend to have simple lance-shaped leaves with parallel veins (see Dicotyledonous).

Photosynthesis. The reaction by which glucose is formed in the chloroplasts of plants from water and carbon dioxide in the presence of chlorophyll, which acts as a catalyst. Oxygen is produced as a by-product.

Retting. Literally means rotting. Flax straw, for example, is pulled and soaked in water (sometimes in ponds) in order that decomposition by fermentation loosens the soft, woody, vegetable material from the strong, inner fibres. The retted straw is then scutched (q.v.) to break up the woody material and remove it without damage to the strong pliant fibres.

Roving. The unspun rope of fibres drawn out in the final process before spinning.

Scutching. The operation of separating the woody part of deseeded or retted flax straw from the fibre.

Shives. Short pieces of woody waste material beaten from flax straw during scutching (see Retting).

Sliver. During the pre-spinning preparatory process fibres are carded and sometimes combed (as with cotton) to make parallel and clean the fibres. Once this has been done the fibre is made into a continuous untwisted rope about 20 mm in diameter, called a sliver. The sliver is drawn out to produce a roving which is then drawn out still further to produce a yarn.

Soft fibre. See Hard fibre.

Straw. Flax in its natural, harvested state before any further processing.

Strick. A small bunch of scutched flax of a size which can be held in the hand.

Strike. A small bunch of jute, usually 1 to 2 kg, but smaller than a head (q.v.).

Staple length. The length of the fibre in spinnable form. Vegetable fibres vary enormously in staple length from a few centimetres in the case of cotton to several metres in the case of abaca.

Tow. After scutching flax or hemp fibres are carded and combed to parallel the fibres. The short fibres which are removed during any of these processes are referred to as tow, and are used to produce thicker, rougher yarns.

JOURNALS AVAILABLE FROM IT PUBLICATIONS

Waterlines
Appropriate technologies for water supply and sanitation

Waterlines is the only journal devoted entirely to low-cost water supply and sanitation. Contains practical help and advice on the problems that face policymakers, water practitioners, engineers and fieldworkers in their work. Also includes news and views from the field, the latest resource materials, and jobs, courses and training opportunities in the water and sanitation sector.

Waterlines is designed for the professional – both specialist and non-specialist – whether administrator or engineer, project manager or policymaker, trainer or fieldworker.

Appropriate Technology
Practical approaches to development for the South

Wherever you are in the world, you will find plenty to interest you in *Appropriate Technology*. By concentrating on real-life experiences and problems, *Appropriate Technology* deals with the issues in practical development in a clear, straightforward way – and the lessons can be applied in every part of the globe.

Every issue is based around a particular theme, with up to eight different articles giving a range of viewpoints from across the development sector as well as an up-to-date resource guide, diary of key events, a technical brief, practical question-and-answer sections, news on the latest technologies and developments throughout the North and South.

Small Enterprise Development
An international journal

Small Enterprise Development provides a forum for those involved in the design and administration of small enterprise development programmes in developing countries.

Detailed articles report original research, programme evaluations and significant new approaches, case studies of small enterprise development projects implemented by donor agencies, short practice notes from the various regions of the world describing programmes in operation and work of wider interest, as well as a regular review of resource materials, forthcoming conferences and events and letters from readers.

Small Enterprise Development tackles the major themes and pressing concerns of small enterprise development. This journal will be an invaluable resource to those involved in the small enterprise development sector.

For information on either how to obtain a sample copy, or how to subscribe to the above journals, please contact: Intermediate Technology Publications, 103-105 Southampton Row, London WC1B 4HH.